EAI/Springer Innovations in Communication and Computing

Series Editor
Imrich Chlamtac, European Alliance for Innovation
Ghent, Belgium

The impact of information technologies is creating a new world yet not fully understood. The extent and speed of economic, life style and social changes already perceived in everyday life is hard to estimate without understanding the technological driving forces behind it. This series presents contributed volumes featuring the latest research and development in the various information engineering technologies that play a key role in this process.
The range of topics, focusing primarily on communications and computing engineering include, but are not limited to, wireless networks; mobile communication; design and learning; gaming; interaction; e-health and pervasive healthcare; energy management; smart grids; internet of things; cognitive radio networks; computation; cloud computing; ubiquitous connectivity, and in mode general smart living, smart cities, Internet of Things and more. The series publishes a combination of expanded papers selected from hosted and sponsored European Alliance for Innovation (EAI) conferences that present cutting edge, global research as well as provide new perspectives on traditional related engineering fields. This content, complemented with open calls for contribution of book titles and individual chapters, together maintain Springer's and EAI's high standards of academic excellence. The audience for the books consists of researchers, industry professionals, advanced level students as well as practitioners in related fields of activity include information and communication specialists, security experts, economists, urban planners, doctors, and in general representatives in all those walks of life affected ad contributing to the information revolution.
Indexing: This series is indexed in Scopus, Ei Compendex, and zbMATH.

About EAI

EAI is a grassroots member organization initiated through cooperation between businesses, public, private and government organizations to address the global challenges of Europe's future competitiveness and link the European Research community with its counterparts around the globe. EAI reaches out to hundreds of thousands of individual subscribers on all continents and collaborates with an institutional member base including Fortune 500 companies, government organizations, and educational institutions, provide a free research and innovation platform.
Through its open free membership model EAI promotes a new research and innovation culture based on collaboration, connectivity and recognition of excellence by community.

More information about this series at http://link.springer.com/series/15427

Mangesh M. Ghonge
Sabyasachi Pramanik • Amol D. Potgantwar
Editors

Software Defined Networking for Ad Hoc Networks

Springer

EAI
RESEARCH MEETS INNOVATION

Editors
Mangesh M. Ghonge
Sandip Institute of Technology
and Research Center
Nashik, Maharashtra, India

Sabyasachi Pramanik
Haldia Institute of Technology
Haldia, West Bengal, India

Amol D. Potgantwar
Sandip Institute of Technology
and Research Center
Nashik, Maharashtra, India

ISSN 2522-8595 ISSN 2522-8609 (electronic)
EAI/Springer Innovations in Communication and Computing
ISBN 978-3-030-91148-5 ISBN 978-3-030-91149-2 (eBook)
https://doi.org/10.1007/978-3-030-91149-2

This Springer imprint is published by the registered company Springer Nature Switzerland AG
The registered company address is: Gewerbestrasse 11, 6330 Cham, Switzerland

Preface

The need for automation and efficient delivery of services is important as the idea of smart cities is emerging. In developing such a concept, ad hoc networks play an influential role in promoting security, comfort, and infotainment services. However, due to high mobility and sporadic access, device heterogeneity presents a number of challenges in the architecture of ad hoc networks. Software-defined networking (SDN) has emerged in this context as a programmable and scalable network that has recently drawn interest from research societies, companies, and industries, both in the management of wired networks and in heterogeneous wireless communication. In the last few years, SDN has initiated a paradigm shift in which centralized systems have been followed by both wired and wireless networks. These architectures are designed to provide an SDN controller responsible for centralized network route selection and dynamic regulation of network behavior.

A variety of similar SDN-based studies on wireless ad hoc networks are covered in this book. It presents a brief on SDN requirements over traditional networking, followed by an elaboration on the fundamental architecture and its layers. Subsequently, in different ad hoc networks such as MANETs and VANETs, SDN applications are defined, along with a focus on evaluating and comparing existing SDN-related research on various parameters. In addition, it covers the effect of the SDN paradigm along with implementation problems in contact with ad hoc networks and examines probable use cases based on the SDN paradigm.

This book offers a comprehensive overview of SDN based ad-hoc networks technologies and exploits recent developments in this domain. It focuses on emerging technologies in SDN-based ad hoc networks. Further, a few practical and innovative applications are present, including network security, smart cities, e-health, and intelligent systems. This book also addresses several key issues in SDN energy-efficient systems, the Internet of Things, big data, cloud computing and virtualization, machine learning, deep learning, and cryptography and its future.

This book provides students, researchers, and practicing engineers with an expert guide to the fundamental concepts, challenges, architecture, applications, and state-of-the-art developments in software-defined networking for ad hoc networks.

Organization of the Book

The book is organized into seven chapters. A brief description of each of the chapters follows:

Chapter "Software Defined Networks: A Brief Overview and Survey of Services" presents software-defined networks, a brief overview, and a survey of services. Chapter "Software Defined Network-Based Vehicular Ad hoc Networks – A Comprehensive Review" reviews software network–based vehicular ad hoc networks. Chapter "Modern Technique for Interactive Communication in LEACH based Adhoc Wireless Sensor Network" deals with the modern technique for interactive communication in LEACH-based ad hoc wireless sensor network. Chapter "Security Challenges in 5G Network" focuses on security challenges in 5G network. Chapter "Software-Defined Networking based Ad-hoc Networks Routing Protocols" discusses SDN-based ad hoc network routing protocols. Chapter "Fuzzy Approach Based Stable Energy Efficient AODV Routing Protocol in Mobile Ad hoc Networks" proposes a fuzzy approach–based stable energy efficient AODV routing protocol in mobile ad hoc networks. Finally, Chapter "Security Approaches to SDN Based Ad-Hoc Wireless Network toward 5G Communication" reviews security approaches to SDN-based ad hoc wireless networks toward 5G communication.

Nashik, Maharashtra, India — Mangesh M. Ghonge
Haldia, West Bengal, India — Sabyasachi Pramanik
Nashik, Maharashtra, India — Amol D. Potgantwar

Acknowledgment

We wish to acknowledge the help of all the people involved in this project and, more specifically, the authors and reviewers that took part in the review process. Without their support, this book would not have become a reality. We thank God for the opportunity to pursue this highly relevant subject at this time, and each of the authors for their collective contributions. My sincere gratitude goes to all the chapter authors who contributed their time and expertise to this book. We wish to acknowledge the valuable contributions of all the peer reviewers regarding their suggestions for improvement of quality, coherence, and content for chapters. Some authors served as referees; we highly appreciate their time and commitment. A successful book publication is the integrated result of more people than those persons granted credit as editor and author.

Nashik, Maharashtra, India — Mangesh M. Ghonge
Haldia, West Bengal, India — Sabyasachi Pramanik
Nashik, Maharashtra, India — Amol D. Potgantwar

Contents

Contributors

Rohit Anand DSEU, G. B. Pant Okhla-1 Campus, New Delhi, India

Shubham Choudhary Department of CSE, M. M. M. University of Technology, Gorakhpur, India

Pankaj Dadheech Computer Science & Engineering, Swami Keshvanand Institute of Technology, Management & Gramothan (SKIT), Jaipur, India

Mohammad Faiz Department of CSE, M. M. M. University of Technology, Gorakhpur, India

Mangesh M. Ghonge Department of Computer Engineering, Sandip Institute of Technology and Research Center, Nashik, India

G. Kirubasri Department of Computer Science and Engineering, Sona College of Technology, Salem, Tamil Nadu, India

Ramchandra S. Mangrulkar Department of Computer Engineering, Dwarkadas Jivanlal Sanghvi College of Engineering, Mumbai, Maharashtra, India

Vipul Narayan Department of CSE, M. M. M. University of Technology, Gorakhpur, India

Vinay Kumar Nassa Department of Computer Science Engineering, South Point Group of Institutions, Sonipat, India

Binay Kumar Pandey Department of Information Technology, College of Technology, Govind Ballabh Pant University of Agriculture and Technology, Pantnagar, Uttarakhand, Uttarkhand, India

Digvijay Pandey Department of Technical Education Kanpur, IET Lucknow, Dr. A.P.J Abdul Kalam Technical University, Lucknow, India

Sanjay S. Pawar Department of EXTC, UMIT, SNDT Women's University, Mumbai, India

Pradeep N Department of Computer Science and Engineering, Bapuji Institute of Engineering and Technology, Davangere, India

Devasis Pradhan Department of Electronics & Communication Engineering, Acharya Institute of Technology, Bangalore, India

Sabyasachi Pramanik Department of CSE, Haldia Institute of Technology, Haldia, India

Rajeswari Department of Electronics & Communication Engineering, Acharya Institute of Technology, Bangalore, India

Gitimayee Sahu Department of EXTC, UMIT, SNDT Women's University, Mumbai, India

Prasanna Kumar Sahu Department of Electrical Engineering, National Institute of Technology, Rourkela, India

Preet Sanghavi Department of Computer Engineering, Dwarkadas Jivanlal Sanghvi College of Engineering, Mumbai, Maharashtra, India

Sheel Sanghvi Department of Computer Engineering, Dwarkadas Jivanlal Sanghvi College of Engineering, Mumbai, Maharashtra, India

S. Sankar Department of Computer Science and Engineering, Sona College of Technology, Salem, Tamil Nadu, India

Jagtar Singh N. C. College of Engineering, Israna, Panipat, India

Hla Myo Tun Department of Electronic Engineering, Faculty of Electrical and Computer Engineering, Yangon Technological University, Yangon, Myanmar

Software-Defined Networks A Brief Overview and Survey of Services

Preet Sanghavi, Sheel Sanghvi, and Ramchandra S. Mangrulkar

1 Introduction

Traditional networks were built during the build time of devices that are outdated today. They enforced policies and implementations that are irrelevant in today's devices. Moreover, they lack the adaptability to modern-day applications that require real-time communications. Under such circumstances, SDN acts as a boon to modern-day network technology as it does not limit dynamic flexibility and helps gratify the demand for heterogeneous applications. Moreover, in the traditional networks, data and control planes are put together in a box with each other thereby causing hardware dependency. Due to this interdependence, variations according to the need cannot be made efficiently, for example, if an administrator wishes to enforce a new logical network plan, then the reconfiguration of all devices directly or indirectly involved is a cumbersome process. In contrast, SDN distinguishes between the control plane and the data plane based on a set of rules using OpenFlow rules which enables efficient usage of SDN on network devices that support OpenFlow protocol standards. With the rise of independence in hardware, previously difficult tasks are solved more easily.

Earlier network environments require a cumbersome amount of edits to be made in case of any changes in the network. On the other hand, SDN-based networks need to fix only the functions of the controller in case of any changes in the network. SDN ensures that the routes are not required to change over and over in the network after deployment.

P. Sanghavi (✉) · S. Sanghvi (✉) · R. S. Mangrulkar
Department of Computer Engineering, Dwarkadas Jivanlal Sanghvi College of Engineering, Mumbai, Maharashtra, India
e-mail: Ramchandra.mangrulkar@djsce.ac.in

M. M. Ghonge et al. (eds.), *Software Defined Networking for Ad Hoc Networks*, EAI/Springer Innovations in Communication and Computing,
https://doi.org/10.1007/978-3-030-91149-2_1

Customers usually find the need to modify the default set of rules by programming to get the desired results. The customers can simply define basic rules before using the service as the network devices are automatically connected. This increases the flexibility to support user engagement and supports the rapid growth of network needs. Customers get access to an enhanced performance by editing the default protocols to increase the flexibility of the system. This demonstrates the ease with which programmers can develop and deploy code in any language flexibly thereby making SDN a cost-effective choice. SDN enables simple deployment of code and encourages real-time parameter changes as well. It provides an increasingly innovative way to manage different communication networks by increasing the ease of development.

SDN has been an extremely impactful domain not only in the education of computer sciences but also in the IT industry. Different research and market predictions indicate that the yearly growth of SDN's turnover will grow to be around 40 percent in a few years. It has also been flagged as one of the top ten technologies to revolutionize the world by MIT researchers. Tech giants like Google and Microsoft and other companies like GAP have been trying to involve SDN-based architectures to augment consumer satisfaction and their profits.

In the surge of such advancements in SDN, one cannot ignore the security threats that come with it. SDN platforms can bring with them a variety of security threats. Denial of service attacks, lack of traffic management, lack of flow management, irregularity and lack of scalability in SDN controllers, inefficient placement of devices, etc. are some of the problems that need to be acknowledged while studying the efficiency of SDN. However, the centralized controller can help in mitigating the threats by using reliable traffic analysis techniques. Moreover, the array of anomaly detection systems involved with SDN can provide important data which can be used to regularly examine the network.

Recent trends are indicators of the fact that many researchers and industrialists have turned their attention to solving the problems related to SDN. These solutions and analyses range from new policies, infrastructures, and automation techniques. However, these contributions are not enough to ameliorate the shortcomings of SDN entirely as compared to private cloud data centers, and further research is expected to aim at increasing the potential of SDN by studying the open challenges in SDN.

Before understanding the infrastructure of SDN in depth, it is necessary for us to understand how SDN evolved over the years and some background information related to OpenFlow and network visualization technologies.

1.1 Evolution of SDN

The development of SDN was massively corroborated by OpenFlow and network function virtualization (NFV). Many concepts of SDN like distinguished control and data planes and requirement-driven tasks were not included in the official architecture due to lack of hardware support. However, after 2008, significant contributions were made to aid in exploiting the potentials of SDN.

1.1.1 OpenFlow

OpenFlow is one of the backbone technologies implemented by the Open Networking Foundation (ONF). It acts as a standard aimed at implementing SDN in a network environment. It builds the protocols for communication between the two most important components of SDN that are the OpenFlow switch and the controller. The switch here serves the purpose of a taxonomy of different packets from a port as well as processing the packets in a particular way based on various packet header fields. OpenFlow forces the controller to guide the switch as to how the data packets should be used. Such guidance is done in the form of flows. Each flow possesses packet match fields, priorities of flows, counters, and a cookie. These flows are categorized into flow tables.

Figure 1 above demonstrates the OpenFlow architecture. It consists of several components such as virtual or physical switches, controllers, and applications. A state graph is maintained of the switches that illustrate a northbound API to the applications by the OpenFlow controller. This northbound API acts as a source of abstraction of the network that enforces the applications to perform certain tasks. These applications are differentiated based on their requirement of strictly OpenFlow switches. Some require a network exclusively of OpenFlow switches, whereas

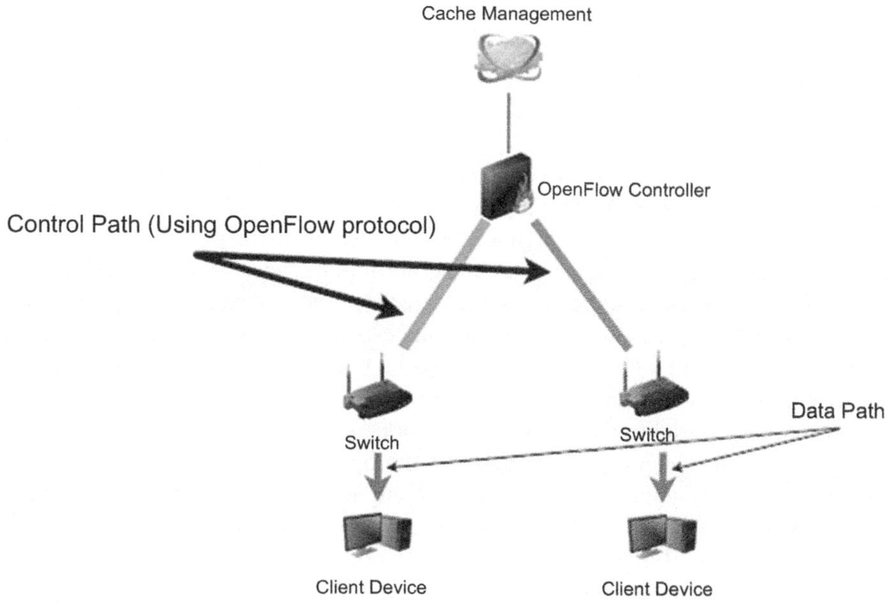

Fig. 1 OpenFlow Architecture

others can adapt to a partial usage of OpenFlow switches also called an overlay or a hybrid network.

1.1.2 Virtualization of Network Components

NFV performs the task of segregating certain functions from the network environment and using them separately as independent software. Functions like firewalls, virtual routing, and traffic control are called virtual network functions that are executed as software in virtual machines. They are mainly used by network administrators to decrease the time needed to deploy updated products in the market and better scaling of resources. They aim to increase the agility of the deployment while keeping the operational and capital costs constant. NFV mainly involves three main features

1. Virtualized function Possesses the software for execution.
2. NFV infrastructure Possesses logic for implementation.
3. Orchestration and management of NFV Enable ease of access, storage, and exchange of information for optimized network performance.

Some other benefits of NFV can be described as follows

1. Centralized data center for increased resource management efficiency.
2. Adaptability to changing business and requirements.
3. The absence of fixed, static hardware decrease the time required to deploy and configure updates.

2 SDN Architecture

Computer networks are built as a concatenation of three different planes forwarding plane, as the name suggests, this plane consists of a variety of internetworking forwarding devices like routers, switches, hubs, etc. which are responsible for forwarding the network traffic. Control plane the control plane hosts the protocols that are responsible for instructing the data plane devices on how to handle and forward the packets. Management plane as the name suggests, this plane is responsible for the overall management of the SDN framework. It hosts different networking and security-related policies that are required for efficient management of the network. When it comes to traditional computer networks, the control and the data plane are firmly integrated. This integration increases the rigidity and the static component of the network. Consequently, the development and deployment of new network applications in the framework become a strenuous task. This is because in order to do that, a hardware modification would be required in the control layer of all internetworking devices. Software-defined networking (SDN) has been designed for the development and deployment of agile, efficient, and cost-effective functioning and

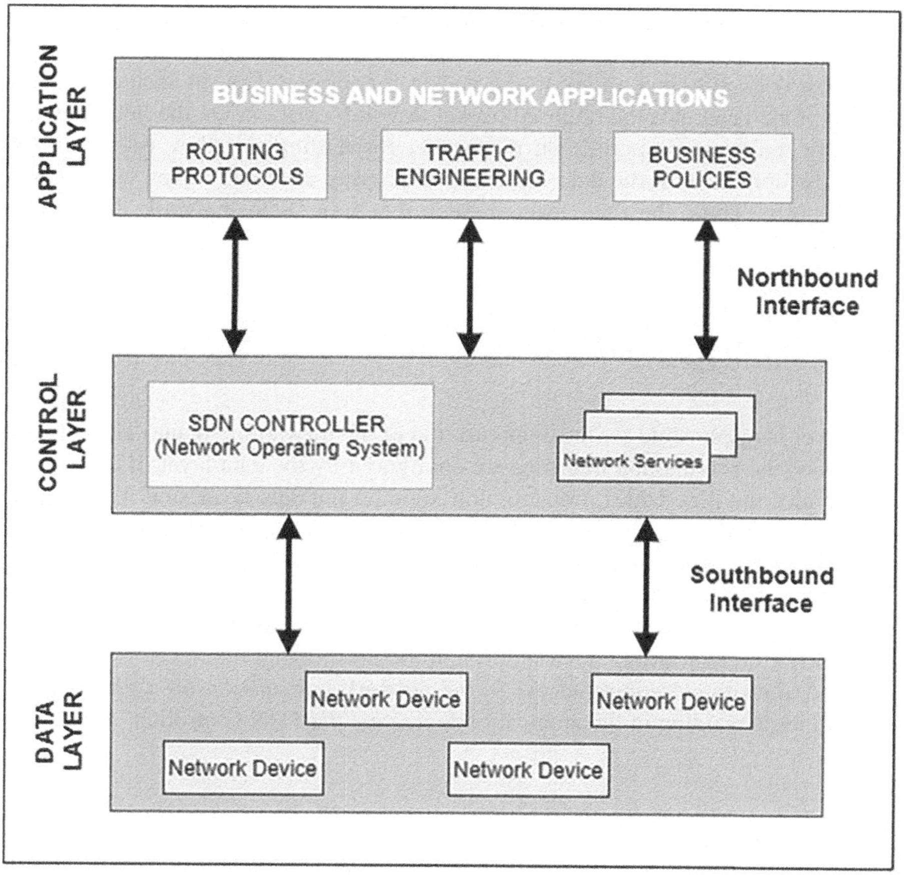

Fig. 2 SDN architecture

framework of a network. SDN is a new networking paradigm that separates the control and the data plane. It achieves so by eliminating the coupled integration between the forwarding mechanism of the data layer and the routing mechanism of the control layer. This enables SDN to develop and deploy network applications based on the current specifications and requirements.

As shown in Fig. 2, the SDN architecture consists of three layers.

2.1 Data Layer

The data layer houses packet forwarding and filtering devices such as routers, firewalls, switches, intrusion detection systems (IDS), computers, and other internetworking devices. Data transmission, the collection or reproduction of information, packet filtering, local network surveillance, and flow statistics are among the main responsibilities of the data layer. The data layer communicates with the control layer if it does not have access to the packet forwarding data. In such a case, the control plane generates the required packet flow information. On the other hand, if there are multicast properties on the packet forwarding rule/flow rule, then the packet is replicated on the data layer before passing different copies via different output ports. Thus, the data layer creates the basic SDN network distribution topology.

2.2 Control Layer

This layer is responsible for the configuration of the forwarding plane. The network devices of the southbound interface are configured by the data layer. It determines and decides the flow tables, transmission logic on the data layer, and its programming. It is also known as a network operating system (NOS) because it functions as the brain of the network. It uses the information provided by the forwarding plane and defines the functioning of the network and its routing logic. This includes single or multiple logically centralized software controllers that are in constant communication with the forwarding devices through the southbound interface. These protocols and algorithms require access to the network information on a global level, relevant logic residing in the application layer, and the SDN controller.

2.3 Application Layer

This layer is home to the network applications that introduce new properties and features such as various new security policies and network performance standards. These features help the control layer configure the network to meet these requirements. The application layer has detailed access to the entire network on a global level which helps it in providing recommendations related to a variety of different application policies and rules. There are two interfaces that separate the layers from each other.

2.4 Southbound Interface

This is an interface for communication between the data and the control layer elements. It specifies the protocols and rules for efficient communication between the data layer devices and the SDN controller.

2.5 Northbound Interface

This is an interface for efficient communication between the application and the control layer elements. It acts as an interface for developers to design applications.

3 Characteristics of SDN

SDN resolves a lot of issues faced when dealing with traditional networks. It has various characteristics and features that help it in simplifying network design and operation. Salient features of SDN that impact its functioning and security are explained below.

3.1 Logically Centralized Control and Network-Wide Visibility

A fundamental feature provided by SDN is a controller that is logically centralized but physically distributed. In the SDN architecture, network control functionality is completely distributed from the forwarding plane via the standard southbound interface. The controller has access to an entire view of the network on a global level and the underlying forwarding information related to the policies that are defined by the services that are running on top of it. NOX [1], Floodlight [2], and Beacon [3] are a few examples of the older controller versions. They functioned as an OpenFlow [2] driver. On the other hand, newer controller implementations like OpenContrail [4] and OpenDaylight [5] provide an update on the required abstractions for the network services. They now support multiple programming interfaces to manage the different forwarding devices.

The devices belonging to the data layer have an active connection to a central controller. The SDN controller sends relevant queries to the data layer devices in order to receive the flow statistics for inferring the network state information. A centralized controller and the overall network status information facilitate decision-making rather than legacy networks where nodes are unaware of the overall network status.

This feature of software-defined networks separates it from today's prevailing networks. Today's networks are built on an autonomous system view, and the network nodes are unaware of the global state of the entire network as a whole. Whereas, in the case of SDN networks, having a global view of the network and centralizing the entire network intelligence fuels the decision-making based on a global network approach.

3.2 Abstraction

Abstraction through the different layers of the SDN network is one of the most prominent characteristics of SDN. In situations where the layers are interfaced using APIs, an SDN setup reduces the load on the programmer. This characteristic of SDN results in the application layer having no direct communication with the hardware elements of the controller. The applications instantiated and deployed in the controller are provided with a detailed view of the network on a global level. Network software and applications can alter the network behavior based on their specifications by using high-level policy languages like Voellmy and Hudak [6]. The SDN controller maps the data layer configuration with these high-level policy configurations. This simplified data layer is very flexible and can add new features to the network.

In an SDN network, the network software resides in the control layer of the SDN architecture. Business applications utilizing SDN services are abstracted from various underlying network technologies. Network devices are abstracted from the control layer which in turn promotes portability and future-proofing of investments in-network services. Frenetic [7] and pyretic [8] are some of the most popular abstraction tools available.

3.3 Network Automation and Dynamicity

SDN offers versatility to handle complex shifts thereby increasing the dynamicity. Depending on the changing network status and conditions, data layer devices can easily be reconfigured. It allows the implementation and deployment of on-demand network and security applications in data centers and the network of service suppliers. In terms of deployment efficiency, this characteristic of SDN drives innovation and allows for the customization of services, flexibility in the overall architecture to adjust to new features and technological changes, as well as a reduction in costs arising from proprietary services. SDN also enables the incorporation of various external network services within its framework. The third-party applications are compiled and run as part of the controller module, while the controller is a monolithic SDN controller. On the other hand, the instantiation of applications takes place without the restart of the controller module in some other implementations of the SDN implementation. The version of the SDN implementation chosen will allow third-party services to communicate with the controller module via internal SPIs or northbound interface-linked APIs (i.e., the REST APIs) that are supported by the SDN controller.

3.4 Virtualization

SDN virtualization requires the sharing and adaptability of physical infrastructures between multiple users in various networks. Virtualizing the components in an SDN framework supports multi-tenancy in the network infrastructure. In a typical SDN network, multiple internetworking devices can be instantiated in a shared physical substrate such that each component represents an individual tenant or customer. Virtualization aims to containerize the SDN components so that individual customer performance, security, and quality of service (QoS) can be personalized for each tenant. Virtualization is of three types storage-based, network-based, and server-based.

The IT community is developing the concept of SDN, while the concept of Network functions virtualization (NFV) belongs to the telecommunication industry. NFV utilizes virtualization software and frameworks so that it can virtualize the network framework that earlier consisted of hardware elements. This leads to the formulation of a network that is dynamic and agile in terms of software provision. VMware, Microsoft, Hyper-V, Citrix, Xen server, and RHEL are some of the most popular companies that provide a network virtualization platform. To actualize virtualization in SDN, a suitable environment must be created where all the tenants of the network can exist together in the same platform while restricting any interference among themselves. Currently, most SDN networks have been using SDNVE by IBM [9] and VMware's Networking Virtualization Proxy (NVP) [10]. NVP can lower down the complexity of the software and programs because of its high-level abstraction feature. SDN on the other hand can handle a myriad of virtual machines very smoothly and easily.

Some common virtualization techniques are as follows

AutoSlice This technique can be found in detail in [11]. It can combat the scalability issue. It can also handle the limitations and constraints related to flow management. FlowVisor (FV) FlowVisor (FV) includes a virtualization scheme for switches that sets up a viable communication link between the data layer and the virtual layer. There is an FV slicer embedded in these switches. They also include an FV classifier. No two FV slicers are managed by the same controller [12]. ADvanced Flowvisor (ADvisor) ADvisor was introduced in order to counter the disadvantages of FlowVisor and provide complete separation and isolation between the different tenants of the SDN network [13]. Carrier grade virtualization This technique is explained in detail in [14]. This technique introduces the inclusion of translation agents in a variety of different internetworking devices belonging to SDN's data layer. A direct communication link is established by the agents between the hardware layer and the individual client controllers without using any hypervisors. Virtualization cloud platform (VCP) This technique depends on the involvement of an operating system within its architecture. In the presence of a network operating system, it can provide proper isolation and efficient utilization of the available network resources [15]. FlowN This technique has been explained in detail in [16]. It can enhance the flexibility of a NOX controller. It is a technique that uses containers to support multiple tenants in the same platform [17]. HyperFlex This technique targets the control layer. It intends to enhance the security, privacy, efficiency, flexibility, and scalability of an SDN setup [18].

3.5 Flow Management

A flow is a basic unit for traffic in the network. A flow rule/flow entry is an entry in a flow table of an SDN switch. The flow table is the primary data structure used by an SDN device. The flow in a network can be controlled by a flow rule which can be subdivided into

1. Flow match fields – used to distinguish different flows using identifiers.
2. Flow priority – used to determine the order in which flow rules will be executed.
3. Flow action – set of operations to either forward the flow or modify it.

SDN switches evaluate incoming packets and take appropriate actions based on the flow table content and packet headers. When data packets arrive at an SDN switch, the switch flow table initiates and executes a flow match process in which it finds all those flow rules that match with the incoming packets. This matching is done in two ways. The incoming packets are compared with the match fields of the flow rules in particular priority order, and then the first match is selected. The matching is done in any order but all matches are recorded. If there are multiple matches, the rule with the highest priority is selected and consequently executed. The selected flow rule is then executed by the network device.

Forwarding decisions in SDN switches is performed for one flow at a time. The basic characteristic of an SDN network is to forward the flow to its controller if the network devices do not have the flow rules required for managing these flows. This enables the flow to be dynamic in nature depending on the various network conditions.

Besides flow-based traffic handling, SDN networks can also use an SDN Flow Manager (SDN-FM) for dynamic flow management. The SDN-FM is used to handle active flows and flow events and manage flow modification requests which include the addition, modification, and deletion of commands. It sends messages to control layer components notifying them about the flow status and flow statistics. Each message contains the unique ID of the flow in its cookie field. Upon successful delivery of the message, the flow is marked as active, and it is added to the active flow list. Each control layer component can send flow modification requests at any time, but only the flow manager can send notification messages to the components. Components can request the deletion of a flow using the unique ID of that flow. If the flow deletion request is accepted, SDN-FM deletes the requested flow, notifies other components about the current flow status, and updates the active flow list. This is how traffic forwarding behavior can be dynamically changed based on network conditions like load balancing, service chaining, and fast rerouting.

3.6 Anomaly Detection

Security in any network is the most important area of concern. In the case of SDN and its centralized control feature, it is comparatively easier to attack the controllers using attacks like DOS (denial of service) attacks, fake packet insertion, running of unauthorized programs on the centralized SDN controller, malicious traffic insertion, etc. Attacks are becoming more advanced nowadays. This makes it very difficult to track the exact source of the attack. Securing the network from such attacks is a challenge. Monitoring the network traffic for different intrusions, surveying and measuring data flow through the SDN network, and implementing efficient intrusion prevention policies are highly crucial to maintain the integrity of the data in our network. For managing the network efficiently, policies for accurate and precise network statistics must be set. This gathered statistical data can be used along with other policies to detect anomalies in the network's traffic flow. Anomaly detection in a network is identifying the events, entries, and observations which do not follow the expected pattern or do not conform to the given rules of the network. Anomaly detection systems are used to analyze packet flow. They are also used to produce warnings about all anomalies for networks. Implementation of anomaly detection policies on the entire traffic in a traditional network structure is very complex, but due to SDN's centralized platform and programmable control layer, they can be easily implemented. In an SDN network, all incoming packets first go to the controller who analyzes each and every packet for any sort of anomalies. It does so using the network management policies and its service integration feature which allows the controller to add and modify policies. SDN networks contain an anomaly detection manager (ADM) [19]. It is responsible for managing a variety of security functions like inspecting packets and detecting any intrusions into the system. It looks out for intrusion alerts and consequently sends details like the intrusion type and information to the CDM [19].

3.7 SMP (Switch Management Protocol)

All the switches belonging to the data layer are used to forward the network traffic and must be programmed via the southbound interface. The SDN controller programs these switches via a standard interface. A companion interface is required to act as a programmable interface for the SDN controller to communicate with the switches in the network. SDN offers the ability to program networks and their components using switch management protocols. There are multiple switch management protocols available such as OF-Config, OVSDB, NETCONF, OpenFlow, etc. [20]. The most common switch management protocol used in SDN networks is OpenFlow. OpenFlow contains a set of switch management protocols that define the functions used to manage the switches in the network using a centralized controller. Using the OF-Config protocol, several logical switches installed on top of the unit can be set and managed. The NETCONF transportation protocol transfers the switch configuration information between the configuration point and the packet forwarding unit. It also specifies the collection of operations over the RPC.

3.8 Open Programmable Interfaces (OPI)

In traditional networks, there is no separation between the control layer and the data. In traditional networks, the control layer and data layer are tightly coupled together. This feature was provided by the SDN architecture. SDN networks have a separation between the said two layers. The main reason why this feature was introduced was to simplify the forwarding devices and allow the network software in the SDN controller to evolve independently. This feature increases the probability of innovation and makes it easy for novel solutions to be incorporated within the network. The applications running in the controller manage the devices belonging to the data layer. It introduces new network and security functions. A standardized programmable interface like OpenFlow can abstract the complexity of network hardware components by programming multiple types of forwarding devices such as network processors, ASIC, virtual switches, etc. There are multiple open programmable interfaces like the following

Control-Data Interface it is often called the southbound interface. Examples are OVSDB, OpenFlow, NETCONF, OF-Config, etc.

Application Control Interface it is often called the northbound interface which includes different REST APIs.

East-West Interface it is present between the SDN controllers. It provides a bidirectional communication interface between SDN controllers which belong to or originate from either separate domains or the same SDN control domain. The APIs that can be found in the east-west interface is explained in detail in [21].

4 SDN Operation

SDN is composed of three major functional components the application plane, switches, and the SDN controller. Figure 3 shows the general operation and functioning of SDN. The SDN operation switches have a data structure called a flow table that collects and stores a set of flow rules. When a switch gets an incoming packet, it tests if any of the flow table entries match with it. In the event of a match, it responds by forwarding the received packet. The switch either discards the packet or transfers the packet to the controller if it is unable to find a match. The controller provides a detailed view of the network as a whole to the network software and applications. It allows the SDN applications to define flows on switches and to manage the network functions responsible for packet forwarding. A single controller has the ability to manage a myriad of networking devices. It calculates the forwarding

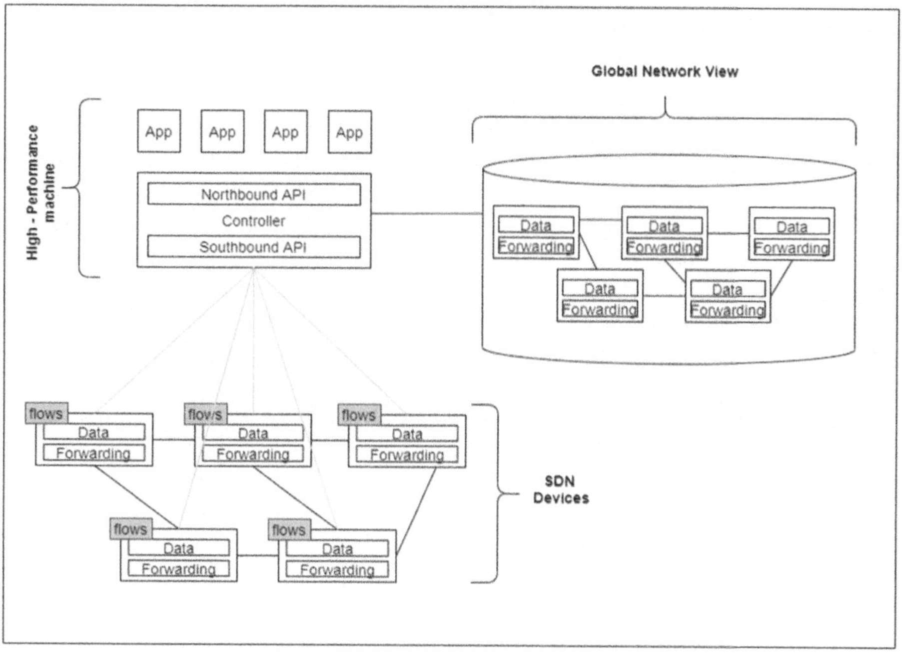

Fig. 3 SDN operation

rules and then operates them using an efficiently performing device that has low latency.

The SDN applications are interfaced with the SDN controller on top of which they are built. This interface assists in receiving packets that have the controller as their destination address. This interface also assists in setting up active packet streams. When the application begins its process, it defines these flow streams, and they continue until the configuration changes. They also set efficient rules and policies for the controller and set new flows in the switches on the basis of the incoming flow of packets in the controller. The switch will now respond locally when it receives an incoming reactive flow. The SDN controller has access to a detailed view of the entire network on a global level. It regulates functions related to traffic management. It also implements different policy decisions like those related to routing, packet forwarding and redirecting, and load balancing. It also controls various internetworking devices belonging to the data plane and provides the applications with an API on the northbound interface of the architecture. It has features like end-user device discovery, network device discovery, network device topology management, and flow management. It contains a variety of software modules that are required for its functioning (as shown in Fig. 3). They are responsible for maintaining the local databases that store the current network topology, flow status, and statistics.

5 Threat Categories in SDN

Security is the most important concern in any network. The security of any network is tested based on how well it can prevent the infiltration of security threats and mitigate them in case of successful infiltration. A threat in the cyber world is a malevolent action that aims to illegitimately get access to a network and exploit it. New cyber threat vectors emerge every day. Consequently, the complexity of the prevention and defense mechanisms required to tackle these issues keeps increasing (Fig. 4).

SDN networks are susceptible to a bunch of threats which can be grouped under various categories like scanning attacks, malware, social engineering attacks, spoofing attacks, network-level DoS attacks, sniffing attacks, web application attacks, and many more. Different threats and the categories they fall under can be seen in Fig. 5 below.

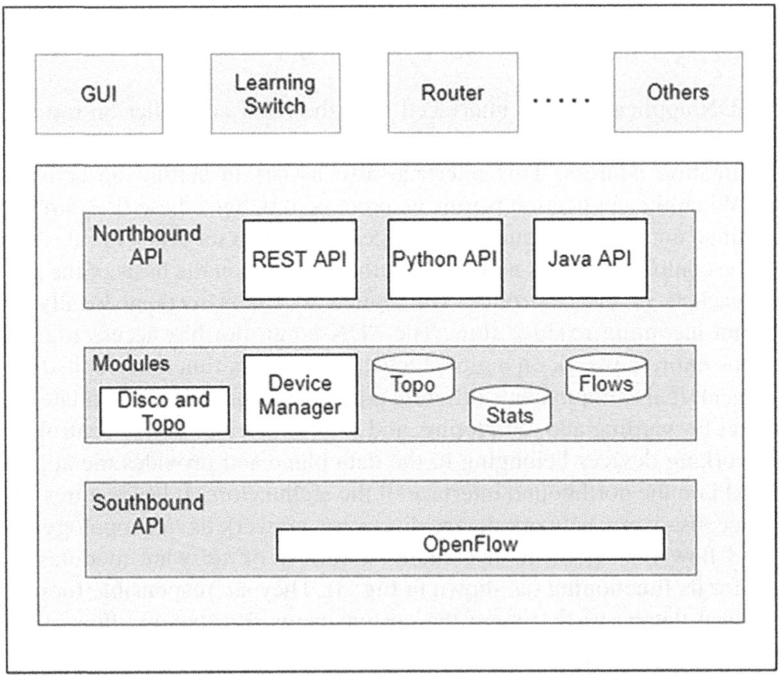

Fig. 4 SDN controller architecture

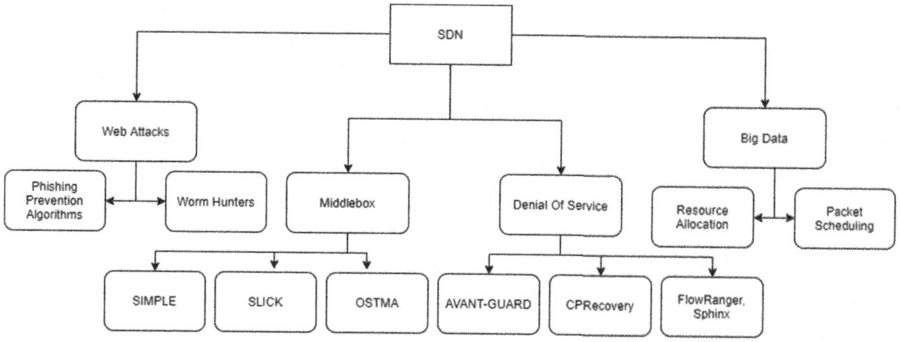

Fig. 5 SDN services

Table 1 Threat categories

Category	Threat categories
Scanning attacks	Network scanning, probing/fingerprinting attack
Spoofing attacks	IP address spoofing, ARP (address resolution protocol) spoofing, network management protocol attack, MAC (media access control) address table overflow, routing spoofing, network management attack, wireless network attack, SSL/TLS attack, a man-in-the-middle attack
Sniffing attacks	Identity information sniffing, covert network channel, information gathering attack, information disclosure attack, credential compromise, eavesdropping, espionage
Network-level DoS attacks	DDoS, UDP flood, ICMP flood, DNS attack, traffic flood, SYN flood, ping of death, teardrop attacks
Web application attacks	Application-level DoS attack, cross-site request forgery (CSRF), XSS, illegal input parameter attack, brute force attacks, misuse of application, injection attack (SQL, command, LDAP injection), authentication, and session management attacks, session hijacking, misconfiguration exploits, brute force attacks, misuse of the application
Malware and social engineering attacks	Ransomware, adware, key logger, malicious scripts, browser attacks, spam, phishing, spyware, virus, spear-phishing, Trojan whaling attack, URL misinterpretation attack, worm
Hardware, operating system, and process attacks	Memory-based attacks, password attack, privilege escalation, chained exploits, buffer overflow, blended threats, kernel-mode exploits, rootkits, process exploits, hardware backdoor, APT

5.1 DDoS Attack Model

Despite the various benefits that SDN networks provide, it does involve certain security problems. Network components like the controller or the switch are quite susceptible to distributed denial of service (DDoS) attacks. A DDoS attack involves the usage of multiple fake host machines that are infused with malicious technologies. These hosts start acquiring and hoarding the network resources (Table 1).

Because of resource hoarding, the network becomes incapable of providing resources to the actual legitimate hosts. The processes running in the controller and its communication capacity overload as attacks by DDoS happen. As a result of all the unnecessary flow generated for the attack packets, the availability of the SDN switch resources goes down. This inadvertently affects the efficiency and performance of the network as a whole. Some examples of DDoS attacks are ping-of-death attacks, teardrop attacks, TCP SYN flood attacks, Smurf attacks, etc. Network systems are exploited on the basis of their weaknesses such as network protocols like IP, TCP, ICMP, etc. The attacker who is using DDoS mechanisms to infiltrate a network may control thousands of computers containing malicious software and use their traffic to consume the target network's resources, bandwidth, and system processing capacity. For example, consider an actual scenario of a DDoS attack where the attacker manages a C&C server and requests connection or transfers malevolent harmful packets to the network under attack by using remote computers.

DDoS attacks can be categorized into two types based on the target of the attack. First is DDoS attacks against the controller, and the second is DDoS attacks against the switch. DDoS attacks against the controller in this case, the attacker tries to occupy controller processes and channel bandwidth between the controller and other parts of the network by flooding the network with the malicious flow. The controller has difficulty differentiating between the traffic from legitimate sources and malicious traffic sent by the attacker. DDoS attacks against the switch in this case, the main target of the attacker is the switch and its flow table which contains flow transmission and control information. At first, the network operationality is tampered with by the attacker by initiating unauthorized access to the network. This unauthorized access could be either physical or virtual methods. Flow tables have limited storage capacity and can only store limited flow rules. Malicious packets are sent by the attacker that comes from an address that the network is not aware of. This leads to the generation and storage of extra rules in the flow table. The flow table's limited capacity fills up due to the entry of these extra flow rules based on the incoming malicious packets. Because of this, the flow table can no longer accommodate any important legitimate flow rules coming from authorized sources. This hampers the network flow.

5.2 IP Spoofing Attack Model

In IP spoofing, an attacker sends malicious packets to the targeted victim. IP spoofing attack involves deceiving and convincing the targeted network that those packets are secure and trustworthy and are authorized to get access to the victim's network/system. The attacker checks the packet stream and then releases malicious packets with forged IP source addresses so that they can conceal their real identities and thereby get access to the network. These malicious packets have the source IP address of a secure and trustworthy host which leads to the network categorizing the packet as legal. IP spoofing is based on the fact that packet forwarding relies only on the packet's destination address and does not verify the authenticity of the packet's source address [22]. IP spoofing allows the attacker to sniff the network in order to identify the network's layout and its secured data using authorized ranges of source IP addresses.

5.3 Drive-by-Download Attack Model

People surfing on the Internet visit a variety of web pages and click on different links. Any of these links could be set by an attacker attempting to use the drive-by-download attack model. Such an attacker intends to insert malicious and insecure scripts in insecure and illegal websites. In such a case, clicking such a link triggers that malicious software and scripts, and they install into the victim's device. It can even redirect the victim to another attacker-controlled illegitimate website. Files containing malevolent source code could download directly to the victim's device. These malicious scripts are implemented during website surfing, viewing emails, or clicking on a pop-up window or ad. If the browser being used has plugins enabled, the attacker can exploit even more vulnerabilities. In this type of attack model, detailed information about the targeted victims or the targeted organization is not needed by the attacker. They only need to be well acquainted with malware, different types of operating systems and web technologies, and cyber infiltration processes.

5.4 Vulnerability Scanner

As the name suggests, a vulnerability scanner develops tools that scan the network, identify potential vulnerabilities and exploit the network. These scanning packets are sent directly to the network by the attacker. The vulnerability scanner begins maliciously exploiting the targeted network if it does not have any defenses against these vulnerabilities. A vulnerability scanner has the potential to gain unauthorized access to the internal networks of its targeted victim organization. The attacker might also have access to the public IP addresses of its targeted victim organization. The attacker requires knowledge of basic or advanced vulnerability scanning tools and knows how to exploit different vulnerabilities in order to pull off such an attack. The attacker might also have knowledge of web technologies, web security, and malicious cyber infiltration processes.

5.5 Malware Controller

The term malware can be defined as any piece of source code, file, application, or software that aims to perform malicious actions such as opening private backdoors illegitimately, deleting important files, falsely behaving as a DDoS agent, downloading other malware, spying, etc. Such malware programs are present in the targeted victim network commune using C&C servers. These remote C&C servers stand for command and control and they are malevolent. The mode of transfer for a malware controller could either be USB disks or online platforms such as websites and emails. If one host in the network is infiltrated with malware, it can spread and propagate to other hosts in the network and connects them to the C&C network. It then executes the malicious commands; accesses private information like keys used in encryption processes, private data and information, and passwords; and sends it to the C&C server. Such an attacker is well acquainted with information regarding the targeted victim network, network protocols, and cyber infiltration processes.

5.6 Phishing Attack Model

In the case of phishing, the attacker uses mediums like emails, SMS, and tweets and falsely showcases them as being sent from a trustworthy source. Such messages sometimes include malicious attachments which when accessed by the victim lead to the installment of malware on the victim's device. Consequently, the malware can then access personal private information from the victim's device. In certain cases of phishing, malicious website links can also be involved. Phishing can be defined as a blend of social engineering and cyberattack tools. The attacker creates malicious messages and emails and false copies of actual legitimate websites to fool the targeted victims into accessing them. A phishing attacker has a good hold over web technology skills, protocols required for email transfer, functioning of different operating systems, malware, and cyber infiltration processes. Such an attack also requires knowledge of common social platforms and websites commonly used by users.

5.7 Eavesdropper Attack Model

Networks have a static nature. This contributes to making it very easy for an attacker to breach the victim's privacy and security using the eavesdropper attack model. The simplicity of an eavesdropper attack makes it an ideal choice for many attackers. As the name suggests, an eavesdropper indicates toward an insider who has access to the private network. The attacker in this case acts as an eavesdropper and has illegitimate access to the private network of its targeted organization. The attacker can access the communication channels, the current status of the network, and private information. There are two modes in which an attacker can operate. In the passive mode, the attacker is present silently in the network and does not send or reply to any packets flowing through the network. Inactive sniffing mode, the attacker uses packet sniffing tools and sends malicious packets to other hosts in the network and its C&C server. If the sensitive data has not been secured with some encryption process, it is collected and stored by the attacker, and it has been encrypted, and then it is collected and stored for later. For pulling off an eavesdropping attack, the attacker must be well acquainted with network protocols, security, and the network architecture.

6 SDN-Based Security Services

6.1 Middlebox Services in SDN Architecture

Middleboxes have been used in different architectures for not only security or load balancing purposes but also for analyzing traffic for attack or irregularity detection. They also help in the examination of the proper framework and methodology required to ameliorate the ramifications of a detected attack. Before the advent of SDN and its features to aid in middlebox services, earlier network operators were required to adopt an intensive and pedantic approach while calibrating different network settings. They had to manually analyze rules for middleboxes and switches to work with the traffic through multiple middleboxes. The advent of SDN in middleboxes revolutionized the way different network settings were calibrated. SDN technology aids by providing an efficient method for steering traffic through a sequence of middleboxes from a central control unit. Multiple systems and architectures like SIMPLE, Slick [23], and FlowTags have been introduced. Slick-Authors of the Slick architecture managed to create a centralized control unit that is able to install and migrate functions into personalized middleboxes. Depending on the security requirements, application interfaces direct the Slick controller to load functions for routing specific logical flows. This increases the flexibility as well as makes the system dynamic as compared to the traditional architectures. Security administrators had to plan the development and deployment along with their locations beforehand which made the system static and void of dynamism thereby making the job more cumbersome for the administrator.

In the Slick framework, the controller assigns the location of the middlebox devices and maintains the ideal pathways for particular traffic to go ahead avoiding any manual labor during the deployment of the device in the location. Using this, only one protocol can be segregated into many parts that can be executed of multiple middleboxes. The Slick controller determines the placement of functions in the middleboxes and establishes the correct paths for specific traffic to pass through those functions removing the requirement to manually plan middlebox placement.

Now, the network administrators can easily implement and deploy rules and policies in the network with the help of a single policy divided into multiple executables that can run on different middleboxes.

Rules for active communication between the administrator and the middleboxes, splitting of the policy into multiple testable fragments, and dynamically enhancing the controller using an algorithm that deploys middlebox functions and routes the traffic were some of the key features of the Slick architecture.

An additional middlebox addition (OSTMA) in [24] is presented for an efficient protective movement with middlebox. The security traversal is aimed at meeting the objectives of the QoS bond of the associated system. The QoS violations are indirectly proportional to the measured delay in the OF controller. Network delay class and the congestion possible in the middlebox attribute to the cost function. The controller and reconfiguration value must be evaluated for a proper estimate of the efficiency period for a particular network. For ensured inspection of crucial network packets by security devices, an SDN application is proposed in [25] that helps restrict network flows. This method is named as CloudWatcher and is designed to work in massive cloud networks. Thus, network applications along with centralized controllers enable relevant protective aid to be employed in network flows. Dynamic configuration followed by combinational defenses and attack detection has made some significant contribution in middlebox deployment.

6.2 Network Security (AAA System)

SDN is driven by the authentication, accounting, and authorization (AAA) system. The AAA system is expected to increase the permanence and accountability of the system. Auth is mainly supported by a Remote Authentication Dial-In User Service (RADIUS) server. This server resides in the application layer and uses the RADIUS server and clients. To ensure and increase the overall strength of the network and to augment its security, an OF controller is implemented with an authentication module. Although the advantage of this OF controller-based access control system is not clearly defined, the AAA functionality is greatly enhanced by the support of SDN. Such a methodology is acquired because of the unevenness and irregularity issues in existing SDN provisions such as OFELIA [26] and GENI [27]. In [118], an interchangeable or transferable certification is issued based on AAA that can be used in any facility.

The authors in [28] aptly identify the impact of solid AAA management mechanisms in order to imbibe SDN's experimentation facilities (EFs). GENI [27] and OFELIA [26] are some of the existing architectures used in EFs, but the problem is with their variability. Umar Toseef et al. proposed an AAA architecture that is based on the protocols of certifications that uses a design understood to be both taut and pliable. The proposed certification is transportable and can be used in any network environment. This certificate can be used for user identification or user authentication to the server for particular services. This solution augments the already existing functionality of AAA and its infrastructure. Although significant advancements have been made in this domain of SDN management, further research is expected.

6.3 Prevention Against DOS Attacks

Different solutions like AVANT-GUARD [29], CPRecovery, Lightweight DDoS, CONA, etc. are presented to prevent the denial-of-service attack (DoS attack). Shin et al. in AVANT-GUARD introduced a solution, called AVANT-GUARD, that includes a connection migration mechanism used to establish useful TCP sessions and actuate triggers that enable data plane devices to activate flow rules under pre-defined conditions. It covers a wide array of problems to solutions under SDN security and its betterment. A connection relocation framework is used to get rid of the poor or crashed TCP sessions before any information is shared with the control plane in AVANT-GUARD. This method ensures that only those flow requested are processed further whose TCP handshakes are completed.

A replication component, CPRecovery [30] is demonstrated for SDN-based networks resilience. CPRecovery acts as a solid backup because it provides a shift from the crashed default controller to a backup controller that is at level with the network's working state.

CPRecovery provides an efficient transition from the failed primary controller to a backup consistent with the network's failure-free state. We proposed another algorithm that is prioritization based for control plane attacks. The author of the FlowRanger [31] managed to examine the trust values for each node in the network in a priority queue that controls the flow rule generation within the administrator. Another solution named Sphinx [32] was proposed by M. Dhawan, in which the information of the controller is studied and relevant messages are detected for creating the network. Different traffic flows like the existing or the earlier traffic flow are segregated on the basis of the period which is then examined and contrasted.

6.4 Prevention Against Web-Based Attacks

Author Masoud in [33] demonstrates that the behavior of users plays an important role in phishing attacks rather than protocols. A pedantic approach using SDN controllers was exuded by implementing a neural network-based phishing prevention algorithm (PPA). Phishing websites mainly aim to use spurious emails or websites designed to fool users into divulging personal financial data by emulating the trusted brands of well-known banks, e-commerce, and credit card companies or for other purposes. Phishing prevention algorithms try to classify website versions as fake or real. If a fake rendition has been detected, then an attempt to redirect to the real website has been made. Other models such as PhishLimiter, an artificial intelligence (AI)-based model, use dynamic deep packet investigation for attack classification.

By calculating the phishing score from the packets, malicious activities are detected with high accuracy of 98.39 percentage. PhishLimiter has two modes of operation fast and slow. In the fast mode, the packet is forwarded to the destination and a copy of it is stored for inspection. In contrast, in the slow mode, all packets wait for the result of the inspection.

A defense system based on SDN was demonstrated in [34] called the WormHunter. It uses multiple honeynet systems under varied network environments in real-time alongside managing flow tables of SDN switches. If irregular traffic is detected, it reroutes it into the net to analyze it more thoroughly. All such data is recorded to train the model in the future.

6.5 Traffic Management

Utilization of the available resources as well as efficient traffic management is one of the key services provided by SDN to uplift the overall network performance. Moreover, to cut back the impact of congestion, traffic engineering is vital. Traditional networks without SDN inclusion make traffic management difficult because it requires manual modifications of rules and makes the full process static in nature. SDN data plane and decoupled control can however help in engineering traffic management policies. Many traffic management and measurement frameworks are published that mainly involve close examination of the network status in real time for managing and controlling the traffic supported different parameters.

Managing traffic is crucial because it aids in cutting out the influence of congestion. However, earlier networks fail to provide dynamicity and require network administrators to manually edit the default foundations in the system for carrying out traffic engineering. Variation in network links or topology can make the work of network operators extremely difficult as it takes a lot of time to converge.

Traffic engineering is extremely important because it helps in reducing the impact of congestion caused by attack traffic or overflooding traffic. In traditional networks, it requires manually modify the foundations or pre-deploy the foundations in network devices for performing traffic engineering. In a succinct language, the heavy interdependence between the data and the control plane negatively impacts the ease with which traffic engineering can be employed in earlier networks and restricts them to be static. SDN can help overcome this obstacle as it provides a dissociate structuring of data and control layers. This decoupling can help network administrators to employ traffic management at the centralized controller.

However, the decoupled control and data plane in SDN can enable network administrators to specific traffic engineering policies at the centralized controller. Additionally, the network managers can increase the utility of SDN in network environments by providing real-time status visualization services. Traffic engineering frameworks mainly consist of two components, namely, traffic measurement and management. Traffic measurement refers to examining and analyzing the network standing instantaneously for managing traffic. End-to-end connection latency, packet loss, etc. are some of the studied network examination parameters.

In 2016, a few researchers proposed a scale restricting method to restrain the amount of traffic that a switch can manage. During execution, if a particular switch cannot engage in traffic management, then another switch that acts as an alternative with a free flow table is used.

6.6 Big Data

Using more than a single SDN controller has enabled researchers to manage large-scale networks.

Several attempts like those of [35] have been made to understand efficient resource and virtual machines allocation using SDN-based cloud centers. In [35], the authors have proposed a new methodology that is relevant to a wide array of applications. Alongside the two algorithms used for resource allocation and power consumption, packet scheduling has been optimized using SDN-based cross point queued switches. The scheduling is refactored based on the type of application. The advantages of this method are as follows

1. Efficient allocation of resources.
2. Reduced power requirements in cloud centers.

Authors in [36] demonstrated SDN-based data centers that involve efficient load balancing to avoid congestion. To avoid congestion, the author managed to frame a method such that the overall load has been detected and distributed along routes that have resources available.

Thus, one can conclude that prominent aspects of SDN like separation of data and control layers; flexible programmability of the network, editions, and reconfiguration of predefined programs for dynamic usage; and an overall dynamic global control can aid in big data processing and data delivery. SDN services can be summarized as shown in Fig. 5.

7 Applications of SDN

A variety of applications and services like security services, network monitoring and intelligence, bandwidth management, content availability, regulation and compliance-bound applications, high-performance applications, distributed application control, and cloud integration can benefit from implementing SDN into their global network.

7.1 Dynamic Storage

Data centers are one of the most commonly used structures for storing information and data. Their layout is complicated and very difficult to handle in traditional networks. The unassigned resources do not have a complete view of the network which in turn leads to inefficient resource utilization. SDN provides the solution to this inefficiency. SDN data centers may dynamically respond to their environments by increasing or reducing their size as required. SDN networks have access to a detailed view of the network on a global level which in turn helps them to efficiently handle traffic, improve the overall efficiency of the network, and reduce energy usage and use of resources. Wide area networks benefit greatly by incorporating SDN in their network. Software-driven wide area networks optimize the network policies and infuse its flow management with the priority levels set by the client. This resulted in a 38-percentage increase in throughput as compared to that of the multilayer protocol switching (MLPS) scheme. SDN can also be used in cellular networks due to their centralized controlled nature. In order to reduce the capital expense and operational costs of cellular networks, SDN offers a variety of cost-effective solutions. Integrating SDN with cellular networking technologies provides an enhanced rate of data transfer and improved quality of service to its customers. SDN-infused cellular networks provide better connectivity and coverage by allowing switching between different wireless technologies. It also improves availability, security, and failure control/recovery in cellular networks.

7.2 Security Services

SDN can also be used in improving security services. Whenever a security violation occurs, every second is very critical to stopping the attack. It is also important to identify the attack and ensure that no other component of the network is infiltrated by the attack. Incorporating NFV into SDN platforms creates a genuinely proactive environment that is capable of risk reduction and responds to security infiltrations promptly.

7.3 Network Monitoring

In network monitoring and intelligence, SDN technologies help in abstracting the network layer within the data centers. Using SDN also helps operators to efficiently manage their bandwidth thereby ensuring that the end users receive optimal bandwidth and network speed. SDN also finds application in regulation and compliance-bound applications where it can control network points, traffic traveling between switches, and hypervisors. It can span various virtualization points, locations, and even cloud locations.

7.4 Blockchain Merger

One of the growing technologies includes blockchain that is proving to be emerging around the analytical and security aspects. Its wide array of uses, mainly focusing and security and maintenance, has helped introduce businesses and market strategists to a more localized, thorough, transparent, and attack-proof solution. Recently, studies involving the merging of SDN and blockchain have shown promising results regarding security architectures. This merging of two growing technologies is possible mainly because of the following points. The advancement brought about by blockchain aids SDN in security purposes. Segregation distributed denial of service attack enforcement and improvements. In the rise of all developments in this area such as improved resource allocation and usage, researchers should simultaneously not be rerouted from the possible attacks that can arise from DDoS as neither of the two technologies is enough by themselves to resolve the issue. More research as promoted in [37] is needed to fully realize the potential and alleviations arising from this merger.

7.5 SDN for Micro Businesses

Given that SDN services are usually employed in wide-reaching environments, there has been some research regarding the usage of SDN in small homes and businesses. Lowly businesses usually involve the usage of cheap or inexpensive materials thereby indicating a greater need for solid security and improved small-scale network organization and maintenance. Moreover, it would be reasonable to assume that a network operator cannot be established in every household or business area. Recent studies have introduced the idea of managing house-enabled networks. In this study, the primary step among is to understand and comprehend exactly the type of data that passes through the network. Moreover, researchers have exclaimed that customers generally find the need to understand deep patterns and trends over their network as they realized that most of the users are not comfortable enforcing policies on earlier networks.

7.6 Optical Network

Multi-network advancements have been made by enabling SDN to expand flows in the form of data traffic thereby augmenting the connection between networks that are packet or circuit-switched. Considering the aforementioned scenario, the Open Networking Foundation also called the ONF originated an optical communication group called the optical transport working group. The advantages of this can be listed as follows

1. Third-party administrations available for deploying.
2. Augmenting the performance of the optical shift stability and jurisdiction of the network environment. Moreover, as seen in [38], Patel and other authors demonstrated an optical connection system integrated with a QoS-enabled singular rule for controlling the switches in an environment assisted by OpenFlow (SDON).
3. Speedy connection dispatch alongside improvised and increased capacity.

8 Potential Vulnerabilities and Open Challenges in SDN

As mentioned before, SDN aids in enhancing the overall network security and increases energy efficiency by providing many solutions such as load balancing and security hardening. However, there are a number of existing open challenges that require further research to fully exploit the advantages of SDN.

8.1 Switch Performance Optimization

One of the two main components of SDN is switches as they contribute to the overall performance of the network. The SDN controller sets up certain rules that are to be followed by the switches. They cannot adapt to the system and follow strictly the logic provided by the controller. One problem with existing SDN switches is that they require high-throughput mainframes with complex logic associated with them. As the complexity of the logic increases, the power consumption of the switches also increases. Thus, further research in this field is expected for developing the existing methodologies related to switches.

8.2 Scalability and Reliability of the SDN Controller

One of the main components of SDN is the controller. It impacts the functioning as well as the performance of the network. Controllers manage up to a limited number of switches after which the problem of load balancing occurs. This can lead to performance degradation as well as decrease the cost-efficiency. Thus, efficient software and infrastructure are expected in novel researches that can help enable agile development and augment the scalability of the architecture in which an optimal number of controllers and their placements should be discussed based on the parameters varied as per need.

Secondly, alarms or exceptions should be raised efficiently to mitigate the problem if there is mishappening of any sort. Under such situations, the reliability of the architecture plays an important role in the software system. Increasing the accessibility of the network to meet the demands of all the sources, understanding the reliability of different production grade SDN controllers, and fault report generation are some of the necessary components to be considered. Additionally, to avoid total failure in times of a crisis, the controller should properly reroute the traffic to an alternative route to maintain flow control and system continuity. Thus, the developers should give increased importance to increasing the reliability of the architecture by developing efficient, secure, and robust control networks.

8.3 Merging SDN and another Traditional System

The nature of SDN and traditional systems vary considerably because SDN systems are mainly flat, whereas earlier ones are stratified in nature. Thus, the integration of the networks because the compatibility and robustness of services and examining systems for both networks are prime concerns. Moreover, SDN is designed to meet the requirements of the customer to provide efficient usage and meet their QoS requirements. On the other hand, traditional services may aim to provide a better experience based on other services. This overlap may degrade the network performance. Thus, an advanced merging technique is necessary to maintain the robustness, compatibility, and reliability of the combined framework.

8.4 Efficient Location of the Device

The placement of different SDN devices like SDN controllers, SDN switches, and others can play an important role in the overall performance of the network. Millions of devices are operated in IoT and big data applications which makes it important to have efficient network resource sharing thereby augmenting the importance of a proper location. Moreover, different SDN architectures work differently as they have different parameters assigned to them. Thus, understanding optimal SDN parameters for the particular use case is also necessary.

8.5 Flow Management

As discussed earlier while examining the importance of SDN switches and their optimization, we realize that SDN switches are not intelligent and they follow a particular set of rules laid down by the SDN controller where the switch tries to map different packets with entries to put forward a calculated judgment. However, the flow tables are restricted to a finite scope, and therefore there is a problem associated with the allocation of the higher number of flow entries. Moreover, as the number of entries increases, the chances of overhead leading also increase which can have a significant impact on the overall performance and energy consumption. Thus, further research is required to devise a method using advanced data structures so that the flow entries can have real-time updates without disrupting the performance or increasing energy consumption.

8.6 Lowly Interface

One of the other problems that have received relatively less importance is the subject of the interface in SDN. Based on our study, there is a growing need to convert advanced and highly organized networks into easy and comprehensible configurations of the switch used. Tasks involving lower levels of intricacy should be able to connect and respond to many other asynchronous events that occur along with the switches.

8.7 Lack of Working Knowledge in Industry

Our research realizes another key hurdle that SDN struggles with, in the large-scale data centers. As observed by Patouni and others in [39], in an absence of complete knowledge regarding the whereabouts and working of SDN, networks can be highly vulnerable to attacks. Moreover, without a proper understanding of how many controllers are required or where the switches are supposed to be placed, SDN cannot be integrated into any establishment or corporation. Moreover, in absence of a designed and organized blueprint, multiple failures in the process of integration can occur. Thus, one can fairly conclude that while software-defined networking holds tremendous unexplored potential, its inclusion in the industry is still in its primitive stages and the technology itself has not quite evolved for existing operators not suited for high complexity. Before enforcing this technology, thorough information and analysis are required related to the architectural and managerial revisions at the operator's end.

9 Conclusion

In this study, we can conclude that SDN is an emerging technology that aids in the management of dynamic networks. We overview different methodologies and architectures that help in solving multiple problems that enforce the advantages of SDN over traditional networks. Moreover, we analyze different open challenges in SDN that need further research in order to demonstrate the utility of SDN more rigorously. Thereby, from our study, we demonstrate a brief overview of cyber menaces categories and possible solutions. We cover solutions for issues ranging from DoS attacks and web attacks to traffic management. We expect that this work will be helpful for researchers as well as students to realize modern defense solutions in SDN-based networking environments.

References

1. N. Gude et al., NOX towards an operating system for networks. ACM SIGCOMM Comput. Commun. Rev. **38**(3), 105–110 (2008)
2. Floodlight Controller, Floodlight Documentation, For Developers, Architecture. [Online]. http//www.projectfloodlight.org/floodlight/
3. D. Erickson, The beacon OpenFlow controller, in *Proceedings in 2nd ACM SIGCOMM Workshop Hot Topics Software Defined Network* (2013), pp. 13–18
4. OpenContrail. [Online]. Available http://opencontrail.org/
5. OpenDaylight. *A Linux Foundation Collaborative Project* (2014). [Online]. Available http://www.opendaylight.org
6. M. Casado, N. Foster, A. Guha, Abstractions for software-defined networks. Commun. ACM **57**(10), 86–95 (2014). https://doi.org/10.1145/2661061.2661063
7. S. Gutz, A. Story, C. Schlesinger, N. Foster, Splendid isolation a slice abstraction for software-defined networks, in *Proceedings of the 1st Workshop on Hot Topics in Software Defined Networks – HotSDN, Helsinki, Finland* (2012), pp. 79–84
8. M. Reitblatt, N. Foster, J. Rexford, D. Walker, Consistent updates for software-defined networks change you can believe in!, in *Proceedings of the 10th ACM Workshop on Hot Topics in Networks, Cambridge, MA, USA* (2011), p. 7
9. C. Dixon et al., Software defined networking to support the software defined environment. IBM J. Res. Develop. **58**(2–3), 1–14 (2014)
10. K. Barr et al., The VMware mobile virtualization platform Is that a hypervisor in your pocket? ACM SIGOPS Oper. Syst. Rev. **44**(4), 124–135 (2010)
11. Z. Bozakov, P. Papadimitriou, AutoSlice automated and scalable slicing for software-defined networks, in *Proceedings of the ACM Conference on CoNEXT Student Workshop, Nice, France* (2012), pp. 3–4
12. R. Sherwood et al., FlowVisor a network virtualization layer, in *OpenFlow Switch Consortium*. (Stanford University, Stanford, 2009). Technical Report. Accessed 1 Aug 2016. [Online]. Available http://archive.openflow.org/downloads/technicalreports/openflow-tr2009-1-flowvisor.pdf
13. E. Salvadori, R. D. Corin, A. Broglio, M. Gerola, Generalizing virtual network topologies in OpenFlow-based networks, in *Proceedings of the IEEE Global Telecommunication Conference (GLOBECOM), Houston, TX, USA* (2011), pp. 1–6

14. P. Skoldstrom, W. John, Implementation and evaluation of a carrier-grade OpenFlow virtualization scheme, in *Proceedings of the 2nd European Workshop Software Defined Network (EWSDN), Berlin, Germany* (2013) pp. 75–80
15. P. Lin, J. Bi, H. Hu, VCP A virtualization cloud platform for SDN intradomain production network, in *Proceedings of the 20th IEEE International Conference on Network Protocols (ICNP), Austin, TX, USA* (2012), pp. 1–2
16. D. Drutskoy, E. Keller, J. Rexford, Scalable network virtualization in software-defined networks. IEEE Internet Comput. **17**(2), 20–27 (2013)
17. J. Reich, C. Monsanto, N. Foster, J. Rexford, D. Walker, Modular SDN programming with pyretic. USENIX Tech. Rep. (2013). Accessed on 1 Aug 2016. [Online]. Available https://www.cs.princeton.edu/âĹjrex/papers/pyretic13.pdf
18. A. Blenk, A. Basta, W. Kellerer, HyperFlex An SDN virtualization architecture with flexible hypervisor function allocation, in *Proceedings of the IFIP/IEEE IM, Ottawa, ON, Canada* (2015), pp. 397–405
19. G. Garg, R. Garg, Detecting anomalies efficiently in SDN using adaptive mechanism. Fifth Int. Conf. Adv. Comput. Commun. Technol. **2015**, 367–370 (2015). https://doi.org/10.1109/ACCT.2015.98
20. ONF Specifications. [Online]. Available https://www.opennetworking.org/sdn-resources/technical-library
21. D. Kreutz et al., Software-defined networking A comprehensive survey. arXiv preprint arXiv1406.0440, 2014
22. C. Zhang et al., Towards a SDN-based integrated architecture for mitigating IP spoofing attack. IEEE Access **6**, 22764–22777 (2018). https://doi.org/10.1109/ACCESS.2017.2785236
23. B. Anwer, T. Benson, N. Feamster, D. Levin, J. Rexford, A slick control plane for network middleboxes, in *Proceedings of the Second ACM SIGCOMM Workshop on Hot Topics in Software Defined Networking (HotSDN '13)*. (Association for Computing Machinery, New York, 2013), pp. 147–148. https://doi.org/10.1145/2491185.2491223
24. Y. Chen, F. Lin, L. Wang, B. Lin, A dynamic security traversal mechanism for providing deterministic delay guarantee in SDN. Proc. IEEE Int. Symp. World Wireless Mob. Multim. Netw. **2014**, 1–6 (2014). https://doi.org/10.1109/WoWMoM.2014.6918983
25. S. Shin, G. Gu, CloudWatcher Network security monitoring using OpenFlow in dynamic cloud networks, in *2012 20th IEEE International Conference on Network Protocols (ICNP)* (2012), pp. 1–6, https://doi.org/10.1109/ICNP.2012.6459946
26. *OFELIA OpenFlow in Europe—Linking Infrastructure and Applications*. [Online]. Available: www.fp7-ofelia.eu
27. *GENI Global Environment for Network Innovation*. [Online]. Available: http://www.geni.net
28. U. Toseef, A. Zaalouk, T. Rothe, M. Broadbent, K. Pentikousis, C-BAS certificate-based AAA for SDN experimental facilities, in *Proceedings of the 3rd EWSDN* (2014), pp. 91–96
29. S. Shin, V. Yegneswaran, P. Porras, G. Gu, AVANT-GUARD scalable and vigilant switch flow management in software-defined networks" in *Proceedings of the ACM SIGSAC Conference on Computer Communication Security* (2013), pp. 413–442
30. P. Fonseca, R. Bennesby, E. Mota, A. Passito, A replication component for resilient OpenFlow-based networking, in *Proceedings of the IEEE NOMS* (2012), pp. 933–939
31. L. Wei, C. Fung, FlowRanger A request prioritizing algorithm for controller dos attacks in software defined networks, in *Next Generation Networking Symposium* (2015), pp. 5254–5259
32. M. Dhawan, R. Poddar, K. Mahajan, V. Mann, Sphinx detecting security attacks in software-defined networks, in *Proceedings of Network and Distributed Systems Security (NDSS)* 2015
33. M. Masoud, Y. Jaradat, A.Q. Ahmad, On tackling social engineering web phishing attacks utilizing software defined networks (SDN) approach, in *2nd International Conference on Open Source Software Computing, OSSCOM 2016* (2016), pp. 1–6. https://doi.org/10.1109/OSSCOM.2016.7863679

34. Y. Hu, K. Zheng, X. Wang, Y. Yang, WORM-HUNTER A worm guard system using software-defined networking. KSII Trans. Internet Inf. Syst. **11**(1), 484–510 (2017). https://doi.org/10.3837/tiis.2017.01.026
35. W. Hong, K. Wang, Y.-H. Hsu, Application-aware resource allocation for SDN-based cloud datacenter, in *Proceedings of the International Conference Cloud Computing Big Data 2013*, (2013–December)
36. Y. Han, Software defined networking-based traffic engineering for data center networks, in *Proceedings of the 16th Asia-Pacific Network Operations and Management Symposium* (2014–September)
37. H. Yang, Y. Liang, Q. Yao, S. Guo, A. Yu, J. Zhang, Blockchain-based secure distributed control for software defined optical networking. China Commun. **16**(6), 42–54 (2019). https://doi.org/10.23919/JCC.2019.06.004
38. A. Patel et al., QoS-aware optical burst switching in OpenFlow based SoftwareDefined optical networks, in *2013 17th International Conference on Optical Networking Design and Modeling (ONDM)* (2013), pp. 275–280
39. E. Patouni, A. Merentitis, P. Panagiotopoulos, A. Glentis, N. Alonistioti. Network virtualisation trends Virtually anything is possible by connecting the unconnected, in *IEEE SDN for Future Networks and Services (SDN4FNS)* (2013)

Software-Defined Network-Based Vehicular Ad Hoc Networks: A Comprehensive Review

Mangesh M. Ghonge and Pradeep N

1 Introduction

VANETs are defined as technologies that take benefit of the new cohort of wireless networks to allow vehicles to communicate with one another over a wireless network. Inter-vehicle communications, improved road safety, emergency warnings, and improved in-vehicle entertainment are just a few of the benefits of VANETs. A growing number of people are becoming interested in VANETs as a result of road transportation problems. Systems of intelligent transportation (ITS) are the primary application of VANETs [1, 2]. Traffic safety and entertainment are the two major categories of ITS applications. A few examples of the former are the following: traffic management, congestion avoidance, routing, data transfer, and traffic signal control [3, 4]. The latter includes Internet access and gaming, among other things. The most critical research in the VANET field is related to developing dependable and efficient vehicle traffic data transmission. Inter-vehicle data transmission may be accomplished in various ways, such as through ad hoc routing, position-based routing, cluster-based routing, broadcast routing, and geocast routing. A growing number of researchers are studying the effectiveness of vehicular ad hoc networks and position-based routing with geocast routing [3, 5, 6] due to their efficiency in applications such as traffic safety. In the current study, the position-based routing method was used to create the route.

M. M. Ghonge (✉)
Department of Computer Engineering, Sandip Institute of Technology and Research Center, Nashik, India

Pradeep N
Department of Computer Science and Engineering, Bapuji Institute of Engineering and Technology, Davangere, India

M. M. Ghonge et al. (eds.), *Software Defined Networking for Ad Hoc Networks*, EAI/Springer Innovations in Communication and Computing,
https://doi.org/10.1007/978-3-030-91149-2_2

Network programmability is a potential technology for simplifying network management while also making it possible for networks to innovate through programmability; SDVN architectures have gained popularity in recent years. As a result, both academia and industry pay close attention to this issue. When control and data planes are decoupled, VANET applications can abstract themselves from the underlying networking infrastructure. SDN technology also provides logically centralized intellect and network state for VANET applications. Merging of SDN and VANET is viewed as a significant direction that can address many of VANET's current problems. To improve user experience, SDN's protruding features such as global topology updates and dynamic management of networking resources were used. In addition to high-throughput and mobility, VANETs require low communication latency, heterogeneity, and scalability among other features. Data centers, cloud computing, and access networks were the primary targets of the SDN paradigm at first. Modern-day networks such as the Internet of Things (IoT) [7], information-centric networking (ICN), and 5G [8] are integrating SDN and OpenFlow techniques. With the Stanford ONRC OpenRoads project as a test bed, [9] found that SDN could support users travel seamlessly between diverse wireless infrastructures [10]. Access points and WiMAX stations can easily switch over to SDN, which also allows video streams to be tricast over both networks (i.e., WiMAX and Wi-Fi). For enterprise WLANs, cloud-medium access control (MAC) [11] provides virtual wireless access points that dramatically improve management. Architectures based on software-defined networking (SDN) are dynamic, sensible, and cost-efficient. Any changes to the network are centrally managed, so they don't require a device-by-device intervention. Cisco says that OpenFlow is the protocol of choice for developing SDN-based applications. The three components of the solution are an SDN controller, an API used for the northbound interface, and an API used for the southbound interface. With OpenFlow, they may provide a single unified view of the entire network. SDN applications such as load balancers, firewalls, and routers all use OpenFlow as the southbound API for sharing information with the controller (Northbound Application Protocol Interface). This allows for the management of the entire system through intelligent coordination, as well as the creation of a framework that allows for the virtualized network and the provisioning of assets as needed. Using SDN as a new network framework, complex network traffic can be solved in a cost-effective and efficient way as a result of the SDN paradigm's applicability, and VANET operations could be changed from what they have been in the past. As a result of SDN's flexibility in meeting the needs of VANETs, it is an appealing approach. With traditional VANETs having difficulty tapping into vehicular technologies, new concepts of SDN integration have evolved. We have decided to collect the latest vehicular network developments for this reason. Researchers in the last few years have integrated SDN functions into VANETs to enable vehicle networks to be used across the globe.

The remainder of the chapter is arranged as follows: the background and overview of SDN and VANET are briefly discussed in Sect. 2. The SDN-based VANETs are described in Sect. 3, and their applications are discussed in Sect. 4. SDN-VANET integration with emerging technologies is described in Sect. 5. Open issues and research directions are discussed in Sect. 6. Section 7 came to a close with its conclusion.

2 Background and Overview

SDN and VANET technologies are briefly described in this section, along with their basic architecture. Our discussion of SDVN architectures in the latter sections of this chapter is limited to the essential details.

2.1 SDN

Figure 1 depicts the main notion of the SDN system. This feature helps the introduction of new routing protocols and management by enabling new routing protocols and management deployment without applying any form of policies or protocols to every device that is connected to the network [13]. OpenFlow, in addition to OpenFlow, is used in SDN (separation of control and forwarding elements). This is certainly the case, but these architectures are designed differently. All of the control and forwarding components are in a single device at CES. The FE and the CNE of

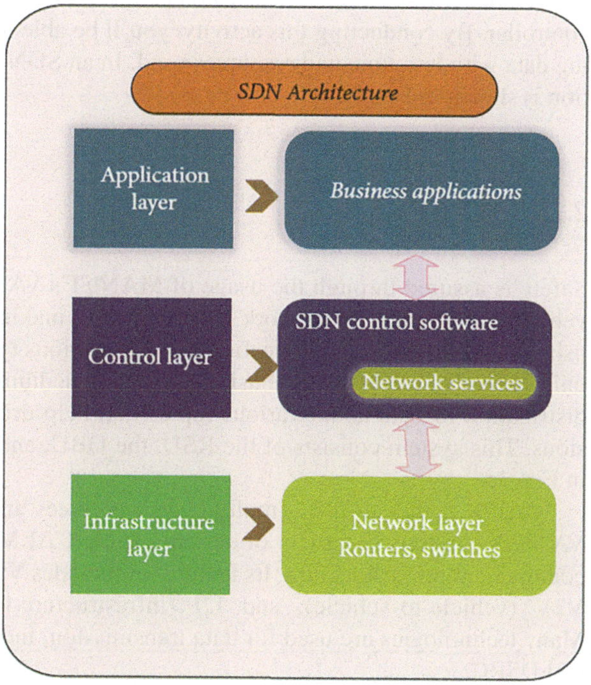

Fig. 1 Basic SDN architecture [12]

ForCES, which are two entities (CE), implement the ForCES protocol by means of the protocol messages it sends and receives. To provide per-packet handling, the FC is used by using hardware to handle each packet [14]. The CE controls the ForCES protocols' functionality. In short, the CE model functions as the head of the process, and the FE one serves as the follower. The use of the logical function block in ForCES is critical (LFB).

This functional block is controlled by the CE via protocols and is dependent on the FE. When the LFB authorizes the CE to configure the FE, the CE can now configure the FE, including configuring how the FE processes the data packet. OpenFlow uses the term "flow tables" to refer to switching devices that have several flow tables in each device. Flow entries providing flow information that indicate how packets are sent are included in this flow table [15, 16]. Receiving packet information is included in Inbound Flow Entries. Data in packet headers is this. The third step, in addition, is to record the packet counts. With OpenFlow, users may make informed decisions about which action to take when a match is identified. The item in the flow tables that the packet refers to is examined, and a certain action is taken if the item is found. OpenFlow uses it to manage the flow of packets throughout the whole network [17]. The OpenFlow system recognizes the traffic flow whenever the traffic flow matches the flow table.

This OpenFlow-enabled system is more effective at using resources efficiently. The network controller makes configuring the system easier because by modifying a single device, the network configuration is modified for all other devices as well. It is not necessary to test network device policies in different network protocols as each device always automatically updates its policies based on the rules set by the controller. By conducting this activity, you'll be able to test and process the receiving data with less time and resources used. In an SDN-based network, communication is shown in Fig. 2.

2.2 *VANET*

Safety is assured through the usage of MANET's VANET field, which serves as a vehicle communication network. The process of making judgments can be aided by using VANET. The communication between various types of vehicles and roadside units (RSUs) is made possible using a wireless medium. This strategy is used for the distribution of data about various topics that help drivers make safe driving decisions. This system consists of the RSU, the OBU, and the AU, which are depicted in Fig. 3.

When it comes to using multiple air interfaces and communication protocols, VANETs could use a variety of options. These CALM services are included in the communication architecture. Its feature set includes V2I (vehicle-to-infrastructure), V2V (vehicle-to-vehicle), and I2I (infrastructure-to-infrastructure) paradigms. Many technologies are used for data transmission, including GSM, WiMax, RFID, and DSRC.

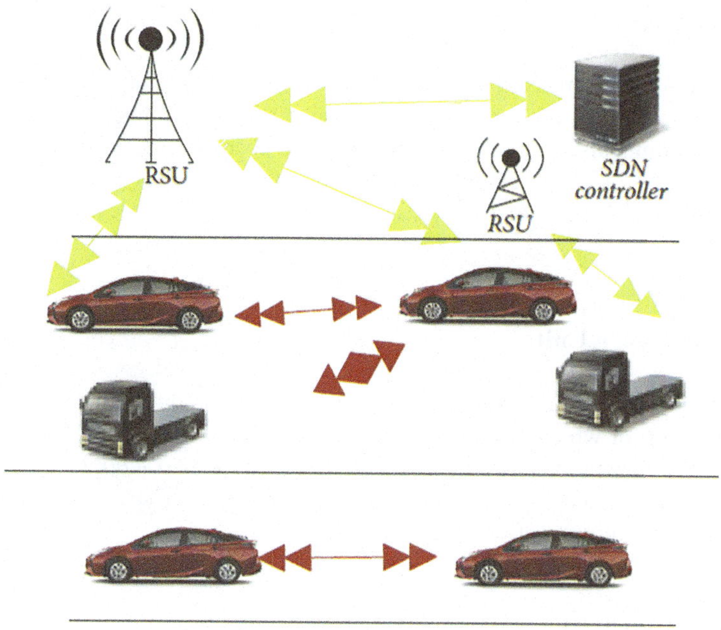

Fig. 2 SDN-based VANET communication design [20]

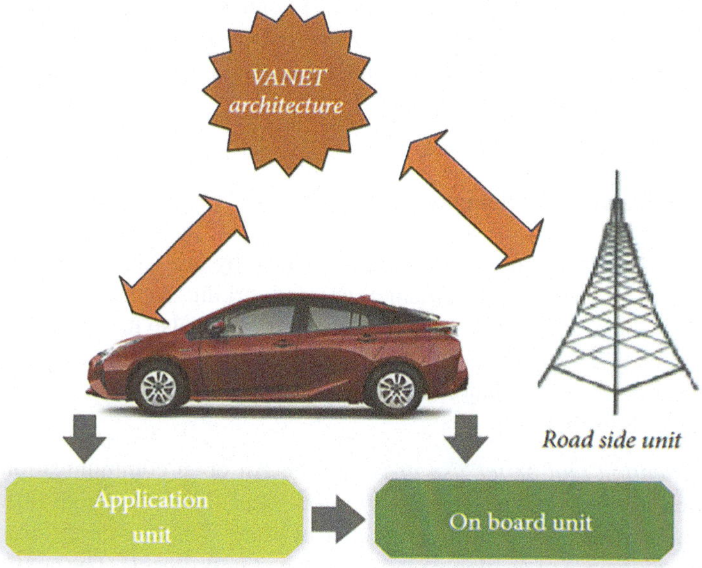

Fig. 3 Architecture of simple VANET [20]

2.2.1 On-Board Unit

On-board units are connected to vehicles and are used to exchange information with other on-board units and reserves to maximize efficiency. The tool has an RSU vehicle communication feature that enables you to read and write data and then sends it to RSUs or other vehicles [18]. A short-range wireless connection is established between the OBU and a user interface and a network device that employs radio technology.

2.2.2 Application Unit

The AU is located within a vehicle and uses the OBU's communication skills. AU and OBU, both of which are situated in the same physical location, are both connected through a wireless/wired connection between the two schools. This personal digital assistant is also good for business applications. There are various OBU devices in AU communication that are in use [6].

2.2.3 Roadside Unit

There is a physical device permanently fixed to the road or parking stand which is known as an RSU. Vehicle-to-vehicle (V2V) communication is provided by RSU devices connected to an Internet source. The RSU is a type of application host and is used to provide service to host an application. The OBU, on the other hand, uses those services to run the application [19].

2.2.4 Benefits and Challenges

The SDN model streamlines network management by unifying network functionality, eliminating obstacles to implementing networking services, and simplifying network management. Despite SDN's many benefits, it also has characteristics that increase security issues, according to a new report.

We've summarized the major advantages and challenges of using SDN technologies in the sections below.

- *Diversification and better use of resources*: An open communication interface provides communication for networking devices that come from different vendors and that are connected to the control plane. Distrust on different componenets and the use of available resources (e.g., OF protocol). The controller takes steps to keep an up-to-date global view of the underlying networking infrastructure at all times. SDNs have trouble being implemented because they can't be reused and can only be used in a single way. Combining multiple physical networks and logical OF-enabled switches, the controller is able to create multiple groups of logical OF-enabled switches, each of which can handle multiple applications and thus appears to be working only for its own application. The instantiation of data plane entities allows the performance of an application to be tailored to its specific requirements.

- *Improved network security*: Communications between the controller and the OF-Switches improve network security. Network traffic analysis and anomaly detection tools enable these switches to collect the required information. As a result of this analysis and correlation, the controller creates or updates a global network view. The entire network can be configured and re-policied based on the results of the analysis. By improving network performance and containing security vulnerabilities, these measures could be beneficial.
- *Single point of failures*: An SDN controller in the middle has a single point of failure, a single central location that is easily attacked, and the low bandwidth connection to the OF-Switch, which is coupled with the flow table size on the OF-Switches. Another negative consequence of networks' support for open programmability is that there is a lack of confidence between data plane entities.
- *Slow propagation of bad information*: The OF-Switch will learn how to process all remaining packets in the flow when a packet belonging to that flow is identified by the OF-Switch. It is not necessary to contact the SDN controller again. Forwarding table rules, however, gets out of sync because of mobility. Packet losses result from a misalignment between the physical and global topologies at the controller. Once the controller modifies the forwarding table entries, however, the topology matches, and no more packet losses occur.

3 SDN-Based Vehicular Networks

People are optimistic about the potential of SDN in terms of restructuring vehicular network infrastructure. SDN has become a reliable approach to network management in recent years. SDN (software-defined networking) uses OpenFlow to interact across the control and data planes. In VANET scenarios, SDN's adaptability works quite well. Today's ad hoc wireless networks are centralized, inflexible, and unprogrammable. One way to loosen constraints imposed by VANETs is to apply SDN ideas to VANETs. VANET networks, built on SDN (software-defined networking), will help to organize networks, provide novel V2V and V2I services, and simplify network management. The program optimizes both the availability of wireless resources and the usage of these resources, including channel allocation and interference avoidance, multi-hop multipath routing, efficient mobility, and network heterogeneity management. Because of how "soft" a concept VANET integration is, it is a highly effective approach (architecture). Since, in the SDN-based VANET architecture, disconnectivity caused by vehicle mobility is mitigated, and the overall connectivity is improved, smart transportation will be supported by this VANET infrastructure. In the sections below, three main aspects of SDN integrated VANET are detailed (Fig. 4).

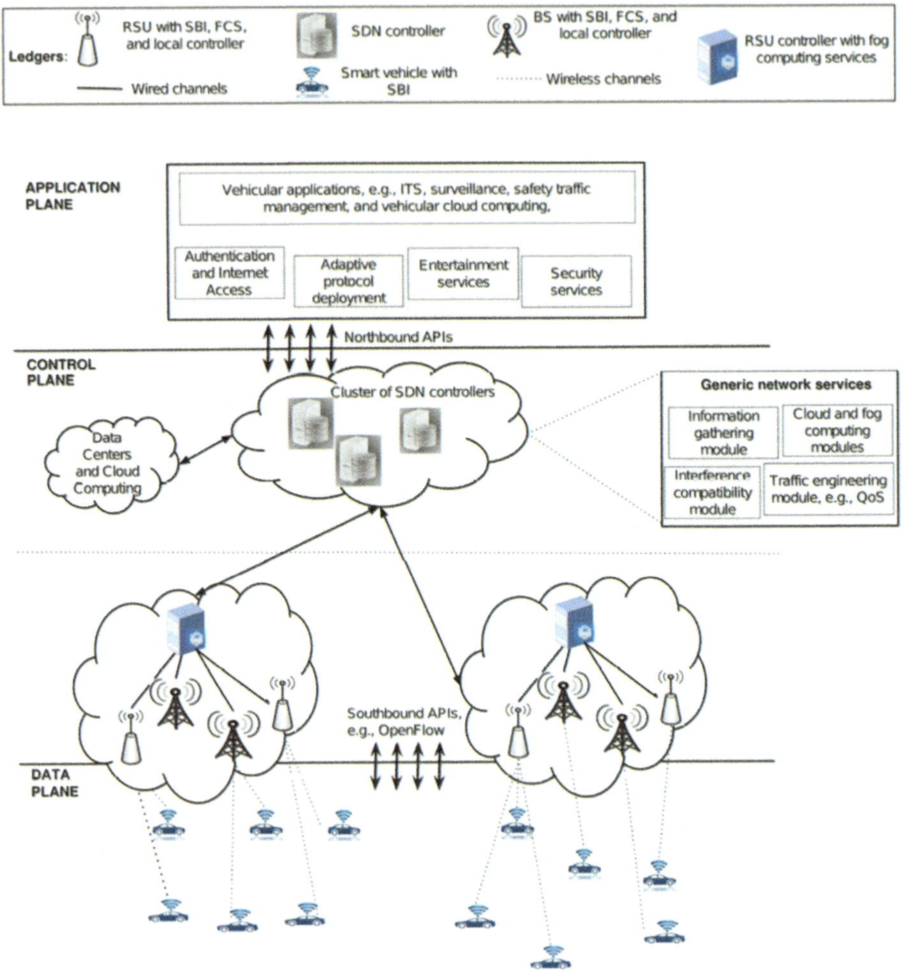

Fig. 4 High-level view of a generic SDVN architecture [20]

- *Appropriate path*: Adding SDN to VANET allows for more detailed routing decisions to be made. In VANET scenarios, data traffic can become congested or slowed down. It occurs when all nodes attempt to use the shortest path, causing some nodes to become extremely congested. As soon as the controller detects one of the scenarios described above, it can stop the process currently running and restart it, increasing the network's efficiency and reducing its blockages.
- *Channel/frequency*: SDN integration in VANET enabled various wireless interfaces or configurable radios to be made available (e.g., cognitive radios). In the case of the radio interface, for example, the controller will make a decision dynamically. It allows for the selection of radio frequencies for different types of traffic at run time. Emergency services are the primary focus of this effort.

- *Transmission power*: Choose the right energy level and transmission range for your wireless interfaces is crucial to the success of your VANET. In a VANET scenario, the controller collects information from vehicles about their immediate surroundings (wireless nodes). A road's emptiness or the distance between vehicles can be determined based on the data collected. To optimize packet delivery, the controller adapts to each node's needs.

A. *Overview of SDVN Architectures*

This section provides an overview of the SDVN research that exists at this time. Future SDVN designs are grouped according to whether they incorporate distinct paradigms (such as 5G, cloud computing, edge computing, and NDN). This SDVN research is outlined in Table 1 with the highlights of the good and bad points of each study.

Following are two broad categories in which we classify the SDVN architectures we surveyed:

- *Developing specific-purpose vehicle networks that are based on SDN*: They are trying to improve VANET aspects, such as network delay, QoS, access control, routing integrity, and security, with the usage of SDN. To increase a given parameter, such as system performance, additional technologies could be required, such as SDN.
- *Developing vehicle networks that are general-purpose based on SDN:* Using SDN and other technologies, such as 5G and named data networking, they strive to increase VANET performance (NDN).

B. *Benefits and Challenges*

As a result of our survey of SDVN architectures presented in the previous section, we have identified several significant benefits and research challenges.

Benefits: It has several advantages, including rapid network configuration, improved user experience due to efficient resource utilization and service latency minimization, as well as resistance to some inherited VANET attacks.

SDVNs have a number of major advantages, which we will discuss below.

- *Optimized resource utilization*: SDVN network administrators have the ability to better manage network resources through the use of the global topology view. When multiple wireless interface or configurable radios are available, for example, controllers can better coordinate channel/frequency selections. He can control both when and how to adjust the wireless signal strength by looking after his company's resources (or network nodes). An SDN controller sends instructions to existing nodes in an RSU coverage area to increase their transmission power because there are too few nodes in that area. SDN controllers can also implement optimal configurations for network devices that are specified by applications running on top. The controller's ability to see the network topology and available resources allows for new options to enhance network performance, allowing for better resource allocation decisions given current network conditions. SDN's demonstrated ability to boost network efficiency can be seen in a study published

Table 1 Review of existing SDVN architecture type

Reference No	Description	Benefits	Drawbacks
[21]	Routing protocol/shortest travel time	The quick provision of packets, as well as low latency and overhead	Maintaining an overview of the global system (or SDN controller)
[22]	SDVN architecture helps increase the flexibility and programmability of the network	Loss of the main controller will have no effect on the improved performance	Not sufficient performance evaluation
[23]	SDVN was made possible by the presence of fog, and autonomous driving overtake was enabled	Accelerate the acceleration of locational awareness and delay-sensitive services	The proposal's usefulness and validity remain unknown, with regard to SDN, VANET, and fog
[24]	delay-minimization routing for SDVNs with mobility prediction	In order to guide trucks on the route	There is no security analysis; thus there will be no remedy if the connection with the controller fails
[25]	Addressing DDoS assaults with SDN-enabled VANET in 5G-enabled VANET	Means that providing a balance between network services, dynamic topology, and network performance and security characteristics are required	Only weak security analysis
[26]	Streaming of multimedia using SDN-enabled buffer-aware VANETs in 5G	Superior quality of service (QoS) during handover	Scalability and robustness in communication
[27]	Throughput and latency improvement provided by fog-assisted SDN-based 5G VANETs	Improving scalability and flexibility is a key benefit to improved communication latency	Out-of-the-box, actual-world, in-the-field performance evaluation
[28]	VANET-assisted 5G, fog-assisted SDN-based VANETs to enhance bandwidth and latency	Using shorter connection latency, scalability, and flexibility as advantages	A demonstration of real-world performance
[29]	An SDN-based 5G-VANET solution that is developed using a social-aware clustering protocol	Promoting better Internet connection performance, decreasing packet delivery	These difficulties are not considered during the test and evaluation of the central controller
[30]	Approaches utilizing a priority-based load balancer are utilized for data-off-loading in SDVNs	Have the ability to scale and control traffic	Evaluation is not done for concerns caused by mobility and security

(continued)

Table 1 (continued)

Reference No	Description	Benefits	Drawbacks
[31]	The use of SDN-enabled MEC architecture in 5G V2V data off-loading	Descriptive route finding using contextual knowledge and V2V off-loading	Knowing where to look for accurate contextual information, privacy when driving
[32]	SDN's unified network resource management strategy provides breakthroughs in vehicle networking technologies	The communication cost is cheap and resources are well-scheduled	If the connection is lost with the controller, there is no solution

in 2016 [33]. In traditional VANETs, any warning message is first sent to the closest RSU for processing. In order to have the message disseminated across the geographic area, the RSU routes the message to the control center, which then broadcasts it to all the other RSUs within the area. Once this has been posted on the Internet, they will then broadcast this message in their respective coverage areas. When it is time to distribute the warning message, a lot of network bandwidth and latency resources are required. Until routes are configured, all future warning messages will go to the controller. Reducing network control overhead, communication latency, and bandwidth consumption are among the many benefits of this change. SDVNs' usage of network resources can improve network performance in certain scenarios, such as both static (e.g., road accidents) and dynamic (e.g., terrorism) (e.g., make way for an ambulance).

- *Quick and versatile network configuration*: Control and logic planes in SDVNs enable rapid and flexible network configuration. Network topology will be able to change to accommodate the mobility of vehicles. Video traffic consumes a lot of bandwidth, and this has congested a few forwarding nodes. Congestion has occurred on a few forwarding nodes due to the path selection method because video applications use a lot of bandwidth on the route. There is an abundance of current network information available at the SDN controller, which enables this situation to be detected and the congested nodes to be identified. In order to avoid the congested nodes, the controller can then reroute traffic around the congested nodes.
- *Heterogeneous network integration*: As a result, heterogeneous networks (wired and wireless), as well as communication technologies (DSRC, Wi-Fi, LTE, 5G, etc.), can be integrated at the data plane in SDVNs. Entities in the data plane and control plane interact more easily when using a communication protocol like OpenFlow. Through the southbound APIs, a data plane switch that supports OpenFlow could, for example, communicate with the controller regardless of the vendor or hardware configuration.
- *Minimizing service latency*: Service latency can be reduced significantly by using SDN at network edge routers. This lessens service latency for delay-sensitive applications. I especially like the fact that SDN's programming

flexibility feature makes it easy to implement fog computing services on edge devices that are SDN-enabled. Routing flow tables can be dynamically reconfigured using SDN global topology information, which helps minimize service latency. As a result, the end user's experience is enhanced. Imagine for a moment that vehicle X leaves the RSU's coverage area. A neighboring vehicle Y could send service messages to C, and Y is within RSU's coverage area. To accommodate the vehicle's growing demands, an RSU could assign more resources to it (i.e., providing support to vehicle X). By utilizing the inherent characteristics of information-centric networking, [34]'s authors provide such services to lost vehicles (ICN).

SDN-based VANETs have many benefits, as was discussed in the previous section.

In this section, we describe the security threats and difficulties associated with SDN-based VANETs. System designers must take security into account when developing the system. There is a brief description of the different types of attackers before moving on.

C. *Attackers in SDN-Based VANET*

A breach could originate inside or outside the organization, may be intentional or unintentional, and could be active or passive. Inside attackers are members of a network and, thus, authenticated, while outside attackers are intruders and, thus, unauthenticated. These outsiders are looking for profit. People who want to do harm to the network for no personal gain only want to do so with ill will. On the other hand, a passive attacker only notices the network is present. With SDN controllers being deployed in networks, the design of an independent communication system is a daunting task [35].

Attacks Against Security Requirements

Hijacking of session: When a session begins, the authentication process is initiated and completed. This can be done easily once the connection has been established. They obtain detailed session information and then act as the central node between the nodes.

Identity revealing: A vehicle's owner in most cases will provide personal information to authenticate the driver. As a result, attackers have an easy time getting into the system.

Location tracking: Using the vehicle's location, you can track it and find out more about the driver and the passengers.

Listening: It targets the network layer, allowing access to confidential information.

Denial-of-service attack: Attacks of this kind are the most common. Nodes are blocked from accessing services by the attackers.

Difficulties in implementing SDN-based VANET: As part of these challenges, there are real-time constraints involved. Next, messages pertaining to security are delivered within a time frame no greater than 100 milliseconds (ms). Real-time constraints require the use of a fast cryptographic algorithm. In order to be effective, authentication must take place in real time. Authentication nodes may engage in malicious activities that may result in accidents, which presents a second challenge. To achieve consistency, a mechanism must be developed. For this reason, the correlation between the data received is examined. An error-prone environment is also a concern for many. On the basis of possibilities, a number of protocols for SDN-based VANETs have been developed. Algorithms used in protocols must take action within a short period of time. Small errors or delays in the algorithm could be harmful. Key distribution is the fourth challenge. On the basis of the key, SDN-based VANET security mechanisms operate. The key is used to encrypt and decrypt the message on the receiver's end. To ensure data security, the key must be protected. Designing security protocols faces a number of challenges, chief among them the distribution of the key. Using SDN-based technology, traditional VANET issues can be addressed. However, implementing the SDN-based VANET poses a number of challenges, including dynamic traffic control and high communication bandwidth [36].

There are a number of major obstacles to implementing a VANET system that is based on SDN.

Traditional networks and SDN communication: In the SDN-based system, policies govern data communication, while in traditional networks there are no policies whatsoever. An SDN-based system should not be deployed in the presence of an existing network since a problem is created. Many classical networks are used to support complicated applications. The SDN-based VANET should be placed on a traditional network with great care to avoid introducing errors. In order for the SDN-based system to function successfully, it must communicate with a traditional network. For this problem, a new routing protocol that specifies the functionalities of the SDN-based system may be developed. This results in a uniform acceptance level. SDN-based systems are being standardized by a task force in Internet engineering [37].

Security issues: SDN-based VANET system security has received scant attention in research. You cannot fully utilize a system if you do not ensure its security. An attacker could exploit the system. Neglecting the security of the SDN-based VANET system allows them to attack the system easily. For the system to run efficiently and be protected from external threats, it must have an effective security mechanism [38].

Availability of Services: In order to keep service availability when a device breaks in a typical network environment, traffic is redirected through an alternate path and processed by another network device. The entire network's operation will be compromised in a centralized system if a device is unable to communicate with another device. Using a standby controller as a backup is an alternative method. Distributed controllers (distributed controllers that use load balancing) can be used to handle this problem as well. The SDN must be able to divert traffic in the event of a network failure by supporting several pathways [39].

Issues related to scalability: There are no differences in the network configuration requirements for routers and switches, which results in the network configuration being straightforward. It is far more complex to configure a dynamic network, as it entails reconfiguring an existing network. In an SDN-based VANET, this is implemented via the control plane. Data traffic increases in direct proportion to the number of traffic on the road. In other words, the greater the traffic on the road, the greater the rate at which vehicles and the RSU (regional services unit) communicate. During periods of excessive traffic, traffic controllers have a hard time maintaining the general flow of the network. Because of the growth in traffic volume, scalability is a big concern. Data flow must be maintained by adding new flow entries to the flow tables. Due to the interconnected nature of all devices connected to the controller, the controller is burdened with an additional processing load. The ideal approach to this problem is to utilize a multicore SDN controller to distribute the workload across multiple cores [40].

4 Applications of SDN-VANET

In this section, we summarized the applications of SDN-VANET

4.1 Application of Comfort Technologies

Most people use comfort applications to get information about the environment, traffic, and the nearest oil stations, hospitals, hotels, and restaurants. Passengers and drivers can exchange messages via the Internet if they have access to it [41].

4.2 Applications for Safety

These types of applications are used to enhance the safety of vehicles, travelers, and drivers. The proposal's sole purpose is to protect drivers and passengers from accidents and to create a transportation-protected space. It collects data from sensors and other vehicles on the road. It is the number of sensors used to collect data and the program used to process it that determines the most important safety and security features [42, 44]. provide examples of this.

4.3 Avoiding Intersection Collisions

It is used at intersections to allow drivers to make choices. When cars are traveling next to it, the RS collects information and processes it in the event of a warning or a mishap. In order for drivers near the changing area to make the proper decision to stop their vehicle, a warning message is sent out [43].

4.4 Stopping Motion as a Warning: Post a Sign

These are used to warn users not to cross the intersection, as there may be dangerous situations in the vicinity of the crossing point. In order to communicate between the RSU and the vehicles sensors, this is necessary. As a result of this program, the driver must pause a couple of times because other vehicles are approaching the intersection in close proximity. It is green-flagged for him to cross after he has passed the turning point [45].

4.5 The Blind Merge Case: A Critical Issue

Using this framework feature, drivers are alerted when visibility at the intersection of two lanes is poor. During the intersection stage, it is used to gather data, determine if the intersection is dangerous, and alert the driver accordingly. Using this application function, drivers are alerted when visibility is poor at the intersection of lanes. It was published in [42].

4.6 Job Areas on High Alert

Using this structure, cars near the work area would receive a warning message to slow down as outlined by [42].

4.7 Cooperative Forward Crash Alert

To prevent collisions with vehicles driving ahead, this feature of the program is activated. In this application function, the V2V form of contact is used to communicate. As a result, you'll know how much risk you're taking by looking at the outcome [45].

4.8 Road Condition Alert (EN)

Optical Biosensing Universal System (OBUS) analyze the sensor and relay study data Results for RSU an alert signal is sent to all vehicles entering a poorly ventilated area so they can stop while sensors are being used to collect information. Application: To keep cars from having to make emergency stops because of an accident, emergency stops are installed on the car [41].

4.9 *Lane Change Warning*

When the driver changes lanes, this application warns him or her because of a too narrow gap between the car in front of him or her and the vehicle in front of him/her. Car-based information is incorporated into this type of technology. As part of this type of alert program, there is V2V communication involved as well [45].

4.10 *Train Ahead of Railway Track*

This system feature alerts the driver when a train is passing in front of him. The contact information of others can be obtained in this way. With respect to your car and the RSU, The RSU transmits an alarm signal to all vehicles in a zone. As outlined by [42].

5 SDN Integration with Emerging Technologies

With augmented reality (AR) applications and a host of others expected to become the new standard, portability and recreation apps will maintain their overall interest. By developing new technologies, it multiplies everything (e.g., IoT, IoV, 5G, FC, and others). As a result, SDN provides the ability to allocate resources to specific flows in a QoS-aware manner while also adapting to traffic changes through powerful network reconfigurations. An overview of some of the technologies is then provided, along with suggestions for future research on SDN to improve at least one aspect of their utility.

A. *SDN And the Internet of Things*

IoT is an important technology when it comes to vehicle networks because it offers new possibilities. Using the Internet of Things (IoT) in tandem with network resources has proven to be helpful [46]. The SDN prototype was implemented alongside IoT in the [47] scenario, which utilized a pilot group of users. To manage dynamic heterogeneous environment in which multiple quality levels are required SDN to be used to achieve a variety of levels for tasks. Besides, because of this blend of technologies, vehicle-related network issues such as intermittent connectivity can't be solved. These technologies must be examined from a broader perspective in order to solve the persistent problems with vehicular networks and achieve effective road traffic management.

B. *Blockchain*

It has been suggested that Bitcoin relies heavily on blockchain technology [48]. Cryptocurrency Bitcoin relies on a key technology called the blockchain. In a blockchain, a cryptographic hashing algorithm is used to create a sequence of blocks that can be viewed by everyone. Such blocks contain transactions involving multiple users, and each new block is associated with a specific miner. The technique of unanimity can be used to select the miner to be used.

As described in [49], a perfect blockchain is maintained by a small group of specified individuals. Using this method, users can keep track of the transactions they've made. As an example, SDVNs [50], 5G [51], and the Internet of Things [52] are implemented using this type of technology. Blockchain technology is explained in [53–57].

C. *Sixth Generation 6G*

Additionally, ML applications such as fast channel equalization and adaptive resource allocation could be used in vehicular networks. One of the primary objectives in developing such a network is to advance intelligent functions of the system while also improving the suitability of the environment for different application requirements, such as ultrareliable and low-latency communications. This network has a three-dimensional space-air-ground structure that gives it an advantage in communicating with humans. SDN allows a more flexible usage of the 5G and 6G technologies, on the other hand. To meet the goals of the 6G objectives, which include NFV, reactive vehicular system control, and cognitive radios, five generations of new methods are proposed, with NFV, reactive vehicular system control, and cognitive radios being the first of these to be employed. Further work needs to be done on intelligent radio, network intelligence, and self-learning with the proactive investigation in order to meet KPIs for the 6G network.

6 Open Issues with Research Direction

Many open questions remain about SDVN's efficiency, scalability, and reliability (trustworthiness), despite its rapid evolution. Some of these challenges could determine the future direction of the SDVN.

- *Management of rapidly changing SDVNs*: Due to the high mobility of nodes (vehicles), the network topology of SDVNs can change rapidly and unexpectedly. Real-time control of vehicles and unstable communication channels is a challenge for SDN controllers or RSUs. Weak DSRC or WAVE connectivity in V2V infrastructure increases the risk of links being broken. This process, however, can be handled with the help of an efficient routing algorithm and infrastructure modifications, but it is costly.
- *Security of SDVN*: A major obstacle to widespread acceptance of SDVN is its security, which is still a major concern. In many SDVN applications, the SDN controller is the primary unit responsible for the overall network's operation. When a single controller is attacked, the entire network can be brought to a standstill. It is possible for a malicious user to gain access to the system and take

decisions in lieu of the controller. Users' safety can be put at risk through such incursions.

- *Latency control in SDVN*: An unsecured wireless connection prevents you from knowing when data will be available. Other aspects of network performance can be optimized in order to reduce latency. There is a clear link between resource optimization and latency control. Because cloud computing is more efficient, it is on the rise. As the number of vehicles in a VANET increase, cloud computing in VANET increases in cost. The total price comes from various functions, such as collecting user information, passing user information to a cloud database, calculating workload in the cloud, providing quality of service (QoS), and tracking location, among others. These various elements all play a role in network latency. With regard to futuristic networks, latency control should therefore be given precedence [50].
- *Scalability of architecture*: As the automobile industry continues to grow, the scalability of existing SDVNs is critical. While traveling, it is impossible to predict whether or not there will be unexpected obstacles or sudden changes. SDVN's performance can be affected by a variety of factors, including technical upgrades, complex road topologies, infrastructure damage, etc. SDVNs with low scalability can also be affected by an increasing number of vehicles and communications.
- *Heterogeneous network*: One of the biggest challenges for an SDN-enabled VANET is bridging the gaps between heterogeneous networks since the future of the SDVN extends beyond vehicle communication. These include new technologies and devices from different manufacturers that have a wide range of features and functions. Due to this, vehicles may be unable to communicate with each other.
- *Evaluation of trustworthiness, detection of misconduct, and revocation process*: The problem of evaluating the trustworthiness of the nodes participating in VANET remains unsolved. Vehicle evaluation mistakes can endanger the lives of users. We still do not have any concrete criteria by which to judge the reliability of any particular vehicle. Although researchers have developed a method for detecting misbehavior [53], there are currently no methods for utilizing this information. For malicious vehicles, there are no specific penalties or fines. No work has been done to implement a revocation process if a node behaves badly in the network.

7 Conclusions

The purpose of this chapter is to give scholars a better understanding of SDN-based VANET systems. This chapter summarizes several SDN-VANET studies in this chapter. Also, it focused on network architectures for vehicles, such as VNI. These architectures are presented in the chapter, with the advantages and disadvantages explained. After that, the dangers and security measures associated with SDNV are discussed. Following that, relevant applications of SDNV and the incorporation of new technologies in SDNV are discussed. Finally, the open issues and the follow-up avenues to research are discussed in detail.

References

1. F. Cunha et al., Data communication in VANETs: protocols, applications and challenges. Ad Hoc Netw. **44**, 90–103 (2016)
2. W.-H. Lee, K.-P. Hwang, W.-B. Wu, An intersection-to-intersection travel time estimation and route suggestion approach using vehicular ad-hoc network. Ad. Hoc. Netw. **43**, 71–81 (2016)
3. A.N. Hassan et al., Inter vehicle distance based connectivity aware routing in vehicular Ad hoc networks. Wirel. Pers. Commun. **98**(1), 33–54 (2018)
4. C. Suthaputchakun, Z. Sun, Routing protocol in intervehicle communication systems: a survey. IEEE Commun. Mag. **49**(12) (2011)
5. N. Noorani, S.A.H. Seno, Routing in VANETs based on intersection using SDN and fog computing, in *2018 8th International Conference on Computer and Knowledge Engineering (ICCKE)* (IEEE, 2018)
6. B.T. Sharef, R.A. Alsaqour, M. Ismail, Vehicular communication ad hoc routing protocols: a survey. J. Netw. Comput. Appl. **40**, 363–396 (2014)
7. S. Bera, S. Misra, A.V. Vasilakos, Software-defined networking for internet of things: A survey. IEEE Internet Things J. **4**(6), 1994–2008 (2017)
8. F.Z. Yousaf, M. Bredel, S. Schaller, F. Schneider, Nfv and sdn 2014; key technology enablers for 5g networks. IEEE J. Select. Areas Commun. **35**(11), 2468–2478 (2017)
9. K.-K. Yap, M. Kobayashi, R. Sherwood, T.-Y. Huang, M. Chan, N. Handigol, N. McKeown, Openroads: empowering research in mobile networks. SIGCOMM Comput. Commun. Rev. **40**(1), 125–126 (2010)
10. M. Kobayashi, S. Seetharaman, G. Parulkar, G. Appenzeller, J. Little, J. van Reijendam, P. Weissmann, N. McKeown, Maturing of openflow and software-defined networking through deployments. Comput. Netw. **61**, 151–175 (2014)
11. J. Vestin, P. Dely, A. Kassler, N. Bayer, H. Einsiedler, C. Peylo, Cloudmac: towards software defined wlans. SIGMOBILE Mob. Comput. Commun. Rev. **16**(4), 42–45 (2013)
12. B.A.A. Nunes, M. Mendonca, X.N. Nguyen, K. Obraczka, T. Turletti, A survey of software-defined networking: past, present, and future of programmable networks. IEEE Commun. Surv. Tutorial. **16**(3) (2014)
13. D. Kreutz, F.M.V. Ramos, P.E. Verissimo, C.E. Rothenberg, S. Azodolmolky, S. Uhlig, Software-defined networking: a comprehensive survey. Network. Int. Architect. **103**(1), 14–76 (2014)
14. H. Farhady, H. Lee, A. Nakao, Software-defined networking: a survey. Comput. Netw. **81**, 79–95 (2015)
15. M. Mousa, A.M. Bahaa-Eldin, M. Sobh, Software defined networking concepts and challenges, in *Proceedings of the 2016 11th International Conference on Computer Engineering & Systems (ICCES '16)* (Cairo, Egypt, 2016), pp. 79–90
16. N. Feamster, J. Rexford, E. Zegura, The road to SDN: an intellectual history of programmable networks. Comput. Commun. Rev. **44**(2), 87–98 (2014)
17. A. Hakiri, A. Gokhale, P. Berthou, D.C. Schmidt, T. Gayraud, Software-defined networking: challenges and research opportunities for future internet. Comput. Network. **75**, 453–471 (2014)
18. S. Al-Sultan, M.M. Al-Doori, A.H. Al-Bayatti, H. Zedan, A comprehensive survey on vehicular Ad Hoc network. J. Netw. Comput. Appl. **37**(1), 380–392 (2014)
19. M. Ealias, R.N. Gaur, A survey on different routing models in cognitive radio ad-hoc network. Int. J. Adv. Res. Elect. Elect. Instrument. Eng. **03**(12), 13741–13748 (2014)

20. W.B. Jaballah, M. Conti, C. Lal, A survey on software-defined VANETs: benefits, challenges, and future directions. arXiv preprint arXiv:1904.04577 (2019) – arxiv.org
21. K.S. Kalupahana Liyanage, M. Ma, P.H.J. Chong, Link stability based optimized routing framework for software defined vehicular networks. IEEE Trans. Veh. Technol., 1–1 (2019)
22. S. Correia, A. Boukerche, R.I. Meneguette, An architecture for hierarchical software-defined vehicular networks. IEEE Commun. Mag. **55**(7), 80–86 (2017)
23. N.B. Truong, G.M. Lee, Y. Ghamri-Doudane, Software defined networking-based vehicular ad hoc network with fog computing, in *2015 IFIP/IEEE International Symposium on Integrated Network Management (IM)* (2015), pp. 1202–1207
24. Y. Tang, N. Cheng, W. Wu, M. Wang, Y. Dai, X. Shen, Delay minimization routing for heterogeneous vanets with machine learning based mobility prediction. IEEE Trans. Veh. Technol., 1–1 (2019)
25. A. Hussein, I. H. Elhajj, A. Chehab, and A. Kayssi, Sdn vanets in 5g: an architecture for resilient security services, in *2017 Fourth International Conference on Software Defined Systems (SDS)* (2017), pp. 67–74
26. C.F. Lai, Y.C. Chang, H.C. Chao, M.S. Hossain, A. Ghoneim, A buffer-aware qos streaming approach for sdn-enabled 5g vehicular networks. IEEE Commun. Mag. **55**(8), 68–73 (2017)
27. X. Ge, Z. Li, S. Li, 5G software defined vehicular networks. IEEE Commun. Mag. **55**(7), 87–93 (2017)
28. A. Soua, S. Tohme, Multi-level sdn with vehicles as fog computing infrastructures: a new integrated architecture for 5g-vanets, in *2018 21st Conference on Innovation in Clouds, Internet and Networks and Workshops (ICIN)* (2018), pp. 1–8
29. W. Qi, Q. Song, X. Wang, L. Guo, Z. Ning, Sdn-enabled social aware clustering in 5g-vanet systems. IEEE Access. **6**, 28213–28224 (2018)
30. G.S. Aujla, R. Chaudhary, N. Kumar, J.J.P.C. Rodrigues, A. Vinel, Data offloading in 5g-enabled software-defined vehicular networks: A stackelberg-game-based approach. IEEE Commun. Mag. **55**(8), 100–108 (2017)
31. C. Huang, M. Chiang, D. Dao, W. Su, S. Xu, H. Zhou, V2v data offloading for cellular network based on the software defined network (sdn) inside mobile edge computing (mec) architecture. IEEE Access **6**, 17741–17755 (2018)
32. Z. He, D. Zhang, J. Liang, Cost-efficient sensory data transmission in heterogeneous software-defined vehicular networks. IEEE Sensors J. **16**(20), 7342–7354 (2016)
33. A. Di Maio, M.R. Palattella, R. Soua, L. Lamorte, X. Vilajosana, J. Alonso-Zarate, T. Engel, Enabling sdn in vanets: what is the impact on security? Sensors **16**(12) (2016)
34. H. Khelifi, S. Luo, B. Nour, S.C. Shah, Security and privacy issues in vehicular named data networks: an overview. Mob. Inform. Syst. **2018**, 1–11 (2018)
35. S. Tomovic, M. Radonjic, M. Pejanovic-Djurisic, I. Radusinovic, Software-defined wireless sensor networks: opportunities and challenges ETF Journal of Electrical Engineering, **21**(1), 74–83 (2015), ISSN 0354-8653
36. S. Sezer, S. Scott-Hayward, P. Chouhan, et al., Are we ready for SDN? Implementation challenges for software-defined networks. IEEE Commun. Mag. **51**(7), 36–43 (2013)
37. H. Kim, N. Feamster, Improving network management with software defined networking. IEEE Commun. Mag. **51**(2), 114–119 (2013)
38. D. Kreutz, F. Ramos, P. Verissimo, Towards secure and dependable software-defined networks, in *Proceedings of the 2nd ACM SIGCOMM Workshop on Hot Topics in Software Defined Networking (HotSDN '13)* (Hong Kong, China, 2013), pp. 55–60
39. A. Tootoonchian, Y. Ganjali, HyperFlow: a distributed control plane for openflow, in *Proceedings of the 2010 Internet Network Management Conference on Research on Enterprise Networking* (2010), p. 3
40. A. Voellmy, J. Wang, Scalable software defined network controllers, in *Proceedings of the ACM SIGCOMM 2012 Conference Applications, Technologies, Architectures, and Protocols for Computer Communication (SIGCOMM '12)* (August 2012), pp. 289–290

41. M. Saini, A. Alelaiwi, A.E.J.A.C.S. Saddik, How close are we to realizing a pragmatic VANET solution? A meta-survey. ACM Comput. Surv. **48**(2), 1–40 (2015)
42. H. Shafiq, R.A. Rehman, B.-S. Kim, Services and security threats in sdn based vanets: a survey. Wireless Commun. Mob. Comput. **2018** (2018)
43. S. Sulaiman, S. Askar, Investigation of the impact of DDoS attack on network efficiency of the University of Zakho. J. Univ. Zakho **3(A)**(2), 275–280 (2015)
44. N. Fares, S. Askar, A novel semi-symmetric encryption algorithm for internet applications. J. Univ. Duhok **19**(1), 1–9 (2016)
45. M. Arif, G. Wang, O. Geman, V.E. Balas, P. Tao, A. Brezulianu, J.J.A.S. Chen, SDN-based VANETs, security attacks, applications, and challenges. Appl. Sci. **10**(9), 3217 (2020)
46. S. Muhuri, D. Das, S. Chakraborty, An automated game theoretic approach for cooperative road traffic management in disaster, in *Proceedings of the IEEE International Symposium on Nanoelectronic Information System (iNIS)*, (December, 2017), pp. 145–150
47. S. Javaid, A. Su_an, S. Pervaiz, M. Tanveer, Smart traffic management system using Internet of Things, in *Proceedings of the 20th International Conference on Advanced Communications Technology (ICACT)* (February, 2018), pp. 393–398
48. S. Nakamoto, Bitcoin: a peer-to-peer electronic cash system. Manubot (2019, November). [Online]. Available: https://www.bitcoin.org
49. E. Androulaki, A. Barger, V. Bortnikov, C. Cachin, K. Christidis, A. De Caro, D. Enyeart, C. Ferris, G. Laventman, Y. Manevich, S. Muralidharan, Hyperledger fabric: a distributed operating system for permissioned blockchains, in *Proceedings of the 13th EuroSys Conference* (2018, April), pp 1–15
50. L. Xie, Y. Ding, H. Yang, X. Wang, Blockchain-based secure and trustworthy internet of things in SDN-enabled 5G-VANETs. IEEE Access **7**, 56656–56666 (2019)
51. B. Nour, A. Ksentini, N. Herbaut, P.A. Frangoudis, H. Moungla, A blockchain-based network slice broker for 5G services. IEEE Netw. Lett. **1**(3), 99–102 (2019)
52. C. Qiu, F.R. Yu, H. Yao, C. Jiang, F. Xu, C. Zhao, Blockchain-based software-de_ned industrial internet of things: a dueling deep Q-learning approach. IEEE Int. Things J. **6**(3), 4627–4639 (2019)
53. M.A. Khan, K. Salah, IoT security: review, blockchain solutions, and open challenges. Future Gener. Comput. Syst. **82**, 395–411 (2018)
54. M. Conti, E. Sandeep Kumar, C. Lal, S. Ruj, A survey on security and privacy issues of bitcoin. IEEE Commun. Surv. Tuts. **20**(4), 3416–3452 (2018)
55. Z. Lu, W. Liu, Q. Wang, G. Qu, Z. Liu, A privacy-preserving trust model based on blockchain for VANETs. IEEE Access **6**, 45655–45664 (2018)
56. F. Tschorsch, B. Scheuermann, Bitcoin and beyond: a technical survey on decentralized digital currencies. IEEE Commun. Surveys Tuts. **18**(3), 2084–2123 (2016)
57. M. Belotti, N. Bozic, G. Pujolle, S. Secci, A vademecum on blockchain technologies: when, which, and how. IEEE Commun. Surveys Tuts. **21**(4), 3796–3838 (2019)

Modern Technique for Interactive Communication in LEACH-Based Ad Hoc Wireless Sensor Network

Rohit Anand, Jagtar Singh, Digvijay Pandey, Binay Kumar Pandey, Vinay Kumar Nassa, and Sabyasachi Pramanik

1 Introduction

Ad hoc wireless sensor networks (WSNs) are usually autonomous and systematic wireless ad hoc networks having a large number of resource constrained sensor nodes [1–3]. An important task of these sensors is organized gathering of data to transfer to a distant base station (BS). Optimizing the energy dissipation for enhancing the lifetime in WSN is of prime importance. An ad hoc wireless sensor network generates a large amount of information that has to be integrated at various levels [4–10]. Bandwidth, memory, signal strength, time, battery power, etc. are the main parameters to judge the performance of a sensor network [11–13].

R. Anand (✉)
DSEU, G. B. Pant Okhla-1 Campus, New Delhi, India

J. Singh
N. C. College of Engineering, Israna, Panipat, India

D. Pandey
Department of Technical Education Kanpur, IET Lucknow, Dr. A.P.J Abdul Kalam Technical University, Lucknow, India

B. K. Pandey
Department of Information Technology, College of Technology, Govind Ballabh Pant University of Agriculture and Technology, Pantnagar, Uttarkhand, India

V. K. Nassa
Department of Computer Science Engineering, South Point Group of Institutions, Sonepat, India

S. Pramanik
Haldia Institute of Technology, Haldia, India

M. M. Ghonge et al. (eds.), *Software Defined Networking for Ad Hoc Networks*, EAI/Springer Innovations in Communication and Computing,
https://doi.org/10.1007/978-3-030-91149-2_3

WSNs are used for a variety of applications like in scientific research, applications related to statistics, fire-tracking applications, defense applications, cognitive networks, healthcare applications, secure routing, etc. [14–20].

Optimization [21–25] is very much important in all the domains of technology. WSN lifetime is an important parameter for the optimum design of data collection schemes for sensor networks. Energy management is a crucial issue in the deployment of ad hoc sensor networks as the sensors depend on the battery for the power that can't be recharged or replaced. The lifetime of these nodes may be increased by considering some special energy-saving scheme.

An ad hoc wireless sensor network consists of spatially distributed autonomous sensors to monitor physical or environmental conditions such as light, sound, humidity, temperature, motion, pressure, and vibration and to cooperatively pass their data through the network to a main location. WSNs pose various challenges such as bandwidth, processing, memory, heterogeneity of nodes, network deployment, power, security, routing and programming, etc.

In ad hoc WSNs, each sensor node is capable to circulate the message through the whole network and combine the message from all the sensors to a sink sensor called as cluster head (CH). It means that all the sensors gather the data from the environment and coordinate with respect to each other by a multi-hop approach. Then, a BS receives all the collected data. The base station is connected to a laptop or a personal computer as shown in Fig. 1. Each node is constituted of the following components:

- Sensor component to perceive the habitat for the information
- Process component to do evaluation from the gathered information
- Communication component to interchange the data with surrounding sensors

Data aggregation refers to the phenomenon of gathering the information from the different nodes to remove the surplus transmission and give obtained information to the base station (BS). Data aggregation usually indicates the integration of information from the multiple nodes (to reduce the traffic) at transitional sensors and transfer of the total information to the BS in energy-saving manner. In the rest of the paper, the term data aggregation will be used to indicate the data collection.

With the help of data aggregation process, robustness and reliability [26] are increased. Also, it minimizes the congestion and saves the sensor energy. But the drawback is that the cluster head may be suffered by the malicious attacker. Another drawback is that the aggregate message may be conveyed to the base station (BS) by a few sensors that results into the more dissipation of energy at these sensors.

There have been so many literatures based on ad hoc wireless sensor network, but most of them are focused toward either energy-saving or life enhancement of wireless sensor networks. Very rarely, the work has been done to improve both the parameters. This motivated the authors to work toward both energy-saving (to improve the communication) and life (durability) enhancement of the wireless sensor network.

In this research, a relay-based work has been implemented in an ad hoc wireless sensor network to increase the network life and improve the network communication.

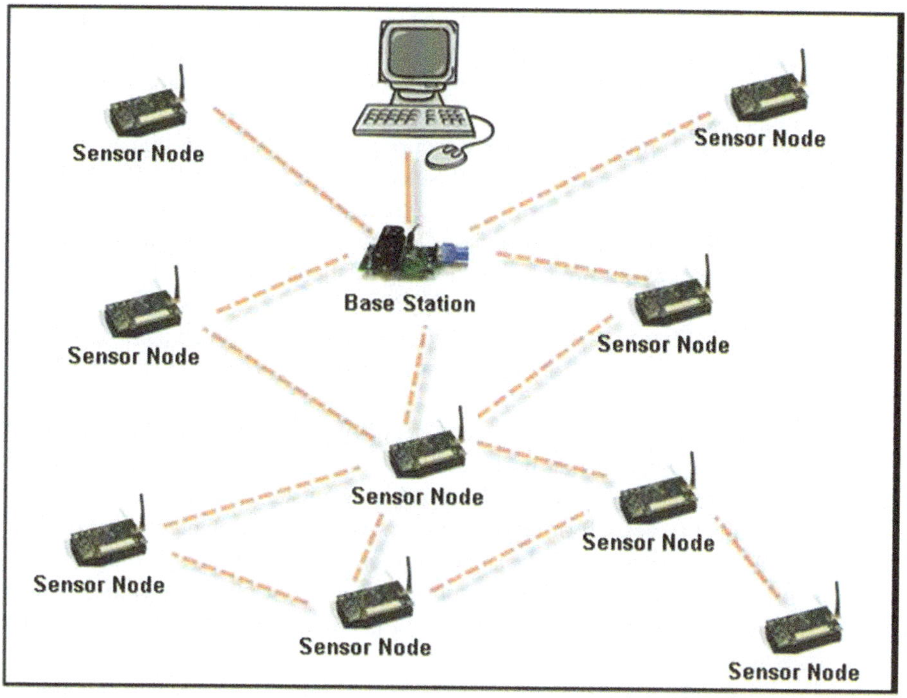

Fig. 1 An ad hoc wireless sensor network

2 Literature Survey

Sohrabi and Pottie [27] proposed an ad hoc wireless sensor network to utilize the bandwidth for saving energy. The method adopted in this paper used incitement of links and scheduling of time procedures. This technique was found to be expandable and fast-connecting as well as auto-organized.

Min et al. [28] proposed a wireless system that has fault tolerance monitoring capability and various applications control on the basis of microsensors operated on different kinds of operating environments. This paper also suggested the different kinds of technologies used for the microsensing like signaling as well as networking based on low power, subdividing of systems, computation based on power awareness, etc.

Manjeshwar and Agrawal [29] proposed a hybrid protocol APTEEN which allows for exhaustive information revival. This type of network helps to retrieve the previous data, current data, and upcoming data from the network. This protocol is able to surpass the other protocols as far as energy dissipation is concerned.

Lin et al. [30] did a survey of commonly used energy-saving medium access control protocols and the various approaches for the ad hoc wireless sensor networks. The requirements for medium access control layer has also been discussed in detail.

Dunlop et al. [31] proposed a time-varying adaptive system that may adapt itself according to the field. A design based on the cross layers was suggested to improve the gain. A smart transceiver design allowed much improvement in the power and distance range.

Chuanhe et al. [32] proposed a highly secure protocol for the mobile ad hoc networks. Security was enhanced by the path selected using a particular theory. In the suggested technique, the mutual trust among the nodes was indicated by two types of trust score. Selection of the path depends upon the trust score. Simulation results proved that a network with some nodes showing misbehavior could also be made to work in a better way.

Varma et al. [33] found that maximum counts of the nodes in an ad hoc wireless sensor network are homogeneous for clustering routing protocols like LEACH, LEACHC, etc.

Yu-quan and Lei [34] proposed better security in LEACH protocol so as to improve the performance of that protocol. The sensing region was partitioned into different equilateral hexagons called clusters. Every cluster is having six equilateral triangles called cells. After forming the clusters, they do not change in all rounds. Each cell and each hexagon have the same number of nodes. In each of the cells, one cell head is selected, and out of the different cell heads, one cluster head is selected. Data is sent in multi-hop form. The proposed protocol was found to have higher energy-saving and higher security.

Kim and Kim [35] suggested an energy-saving scheme in ad hoc wireless sensor network based on the allocation of sensors. With the bound of signal to interference ratio, efficient energy techniques are introduced in the paper based on the adaptive transmit power. Simulation results show that maximum energy-saving may be achieved for a specified bandwidth of the channel.

Berger et al. [36] presented a modified wireless sensor network using a star topology based on time division multiple access. The technique was found to be self-organized by employing the repeater nodes. The transmission area was found to be increased by two times, whereas the power dissipation was found to be average.

Gagneja and Nygard [37] showed that a heterogeneous topology is much more secure than a homogeneous ad hoc wireless sensor network to secure the data. A method called "Improved Tree Routing" was used for routing the data in the network. Through simulation, it was shown that Voronoi-Tabu clustering technique in combination with Secure Improved Tree Routing performs excellently. The suggested method was found to have high throughput and less delay of the network.

Zhang et al. [38] discussed the performance of cognitive vehicular networks for the multi-hop networks. The technique based on relay scheduling showed that the energy-saving and outage probability might be improved.

Ababou et al. [39] proposed an approach that considers the energy level of the nodes which act as relays to indicate whether participation in the transfer of messages is possible or not. The concept of artificial bees has been considered to find the relay node energy during the destination encounter. The suggested approach was found to be energy-saving and lifetime improving.

Andleeb et al. [40] proposed a strategy called mobile sinks to enhance the lifetime of ad hoc sensor networks for the LEACH clustering protocol and TEEN hierarchical protocol. The analysis was done based on the parameters like count of existing nodes, number of inert nodes, remaining energy in the network, etc.

Li et al. [41] described an optimum technique to diminish the loss of information packets in wireless sensor networks with the assumption that the entire state of each sensor is already known by the BS. The results prove that the technique is very efficient as far as network performance and deprivation of packets are concerned.

Ngangbam et al. [42] proposed the organization of the selection of cluster head more appropriately to increase the lifetime of the network with a specific protocol. The output was shown to be better than with the older protocols.

Banerjee and Madhumathy [43] suggested a routing protocol based upon agent cluster that splits the cluster in the various subgroups that make the communication to the cluster head. The routing technique is chosen to reduce the consumption of energy. The suggested method is found to increase the reliability of the network.

Zhang [44] proposed an algorithm in IoT based upon agriculture. Subsequently, a new routing protocol has been proposed to maintain the routing efficiency. The energy consumption of the mentioned technique is lesser than many other previous techniques.

In this paper, an attempt has been done to save energy as well as to increase the durability of ad hoc wireless sensor network based on LEACH protocol.

3 LEACH Protocol

The different kind of rules and regulations for routing used in the network are useful for the safe and energy-saving transmission between the different nodes, between the CH and sensors nodes, and between the CH and BS, etc.

Protocols may be categorized depending upon the infrastructure or protocol functioning of the network (Fig. 2). The protocols [45] that fall in the category based upon protocol functioning are:

- Flat routing.
- Hierarchical (based on clusters) routing.
- Location routing.

The discussed work is based on a specific hierarchical-based routing protocol "LEACH." In hierarchical-based protocols, there is a cluster head which combines the information taken out of other sensors to transfer to the base station (BS). The main intention of using the hierarchical-based rules and regulations is to save energy consumption in sensor networks for a specific cluster. Following are the protocols using hierarchical network model [46]:

- Low-Energy Adaptive Clustering Hierarchy (LEACH) [47].
- Power-Efficient Gathering in Sensor Information Systems (PEGASIS) [48].

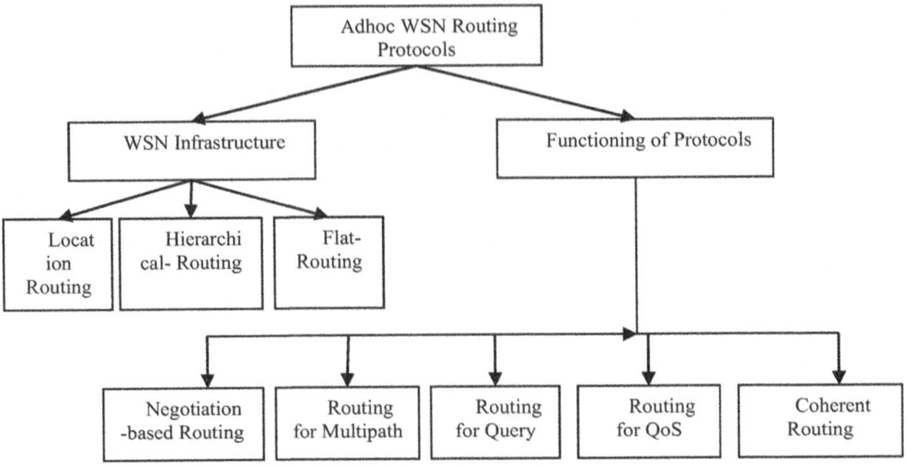

Fig. 2 Classification of routing protocols for ad hoc wireless sensor networks

- Threshold-Sensitive Energy Efficient Sensor Network (TEEN) [49].
- Adaptive Threshold-Sensitive Energy-Efficient Sensor Network (APTEEN) [15].

The LEACH protocol [50] uses a recurrent distributed clustering function to keep the equilibrium energy costs within the entire network. Total time is subdivided into the different rounds, and each sensor node has a definite opportunity of choosing itself as a cluster head. LEACH protocol has the following characteristics:

- Adaptive and self-organized cluster formation
- Localized data transfer control
- Data aggregation.

In that protocol, each cluster head is accountable for making and handling a schedule as per TDMA and thereafter transmitting the combined information from the sensors to the BS which processes using the technique of CDMA [16, 50–52]. This protocol may be subdivided into the different rounds with every round having two phases.

(a) *Setup Phase*: It has two sub-phases

    1. Advertisement phase
    2. Cluster setup phase

(b) *Steady State Phase*: It has two sub-phases

    1. Creation of time plan
    2. Transfer of data

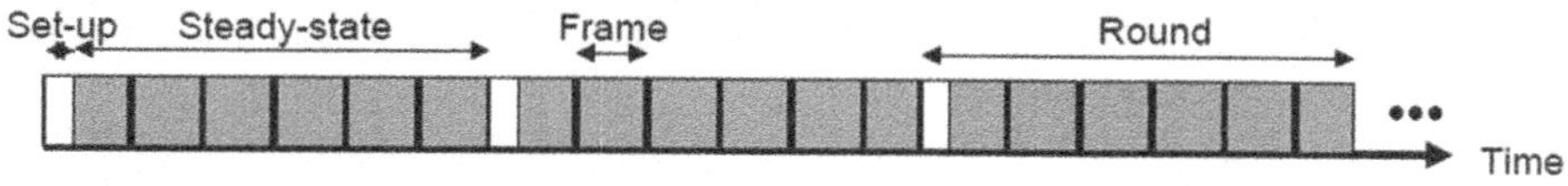

Fig. 3 Different phases of LEACH protocol

Figure 3 indicates all the phases of LEACH protocol.

(a) *Setup Phase*

In the first (i.e., advertisement) phase, the different cluster heads (CHs) notify their surrounding sensors with a specific packet called advertisement packet that they can turn out to be the CHs. The sensors that are not CHs retrieve the advertisement packet with the highest intensity.

In the upcoming second phase, the different sensors instruct the CH with "join packet" that they can turn out to be the members of the cluster. The "join packet" consists of their identities using Carrier Sense Multiple Access. Subsequently, the CH is able to know the count of sensor members and their identities. Depending upon the information gained, the CH makes a schedule as per time slot-based multiple access technique (i.e., TDMA), picks a code of Carrier Sense Multiple Access arbitrarily, and transmits that multiple access table to all the sensors of cluster.

(b) *Steady State Phase*

In this phase, transference of data just starts. All the sensors transmit the information within the assigned slot of time division multiple access to the cluster head. The communication utilizes least energy (dependent upon the obtained intensity of the cluster head advertisement). Every member (i.e., not a CH)'s radio may be shut down unless the sensors are issued the time division multiple access slot, hence reducing the dissipation of energy in all the sensors. After the reception of entire data, the CH combines the entire data and transfers it to BS.

4 Implementation

The present research is done to enhance the network like. In this work, the energy efficiency [53, 54] is increased for the interactive communication in clustered network of high density. The presented work is able to do the interactive communication between the cluster heads. The interactive transmission is done with respect to CHs (cluster heads) directly or with inclusion of relay heads.

The ideas and details of improvement done in the present work are:

(a) In the existing work, the cluster head node is selected randomly in every round. So, a few nodes consume energy earlier because of being acting as cluster head. The norm of choosing a CH (cluster head) protocol arbitrarily picks a CH at one and all round. Hence, a few sensors may consume energy so abruptly. In the present approach, the updated regulation results into the sensors having high amount of residual energy to act as cluster heads. Hence, this increases the long lastingness of the ad hoc network.

(b) In the existing research, energy dissipation between base station (BS) and cluster head (CH) is higher than energy dissipation between the cluster heads resulting in the exhaustion of energy earlier. In this research work, multi-hop communication eliminates the feasibility of dying of ad hoc sensor network rapidly.

(c) In this research work, the choice of relay nodes is done over the network so as to have the interactive communication.

The complete sequence of the proposed technique is indicated in Fig. 4.

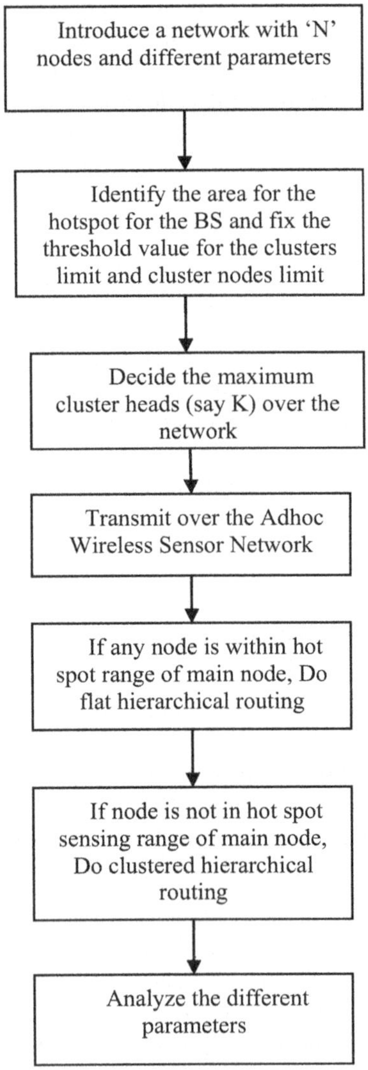

Fig. 4 Flow chart of the proposed technique

5 Results and Discussions

Longevity of an ad hoc wireless sensor network longevity refers to the count of active (alive) nodes, count of inert (nonliving) nodes, packet transmission speed, and duration for which cluster is formed in that network and reduction of energy dissipation in the network. The proposed system gives good performance in all the five aspects, i.e., lesser count of inert sensors and larger count of living sensors are found in suggested technique. Also, energy wastage is reduced and the packet transmission speed is increased. The process of formation of cluster is longer for sure that minimizes the energy loss resulting in the long lastingness of that network.

The cluster head (CH) is totally accountable for managing the communication within a cluster as well with a base station (BS). The CH takes the data from the members of that cluster and then sends that to the BS by straight communication or by interoperative transmission.

This section shows the analysis of the proposed technique with the existing technique with cluster head assumed to be static. This analysis has been done for the different parameters:

- Inert nodes
- Active nodes
- Packet count communicated to BS (without aggregation)
- Packet count communicated to BS (with aggregation)
- Number of packets transmitted in wireless sensor network
- Energy remaining

(a) *Inert nodes*

Figure 5 shows the inert (i.e., inactive) nodes in wireless sensor network in the existing approach. The graph indicates the count of nonexistent (i.e., nonliving) nodes. It can be seen in the existing LEACH protocol approach that in the beginning, the nodes dissipate energy rapidly, and, hence, the nodes become dead at the earlier stage. The complete network gets dead after reaching 380 rounds.

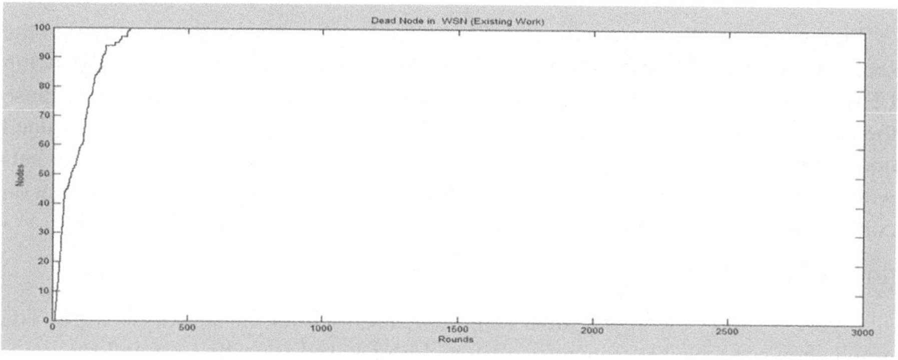

Fig. 5 Inert node analysis (existing approach)

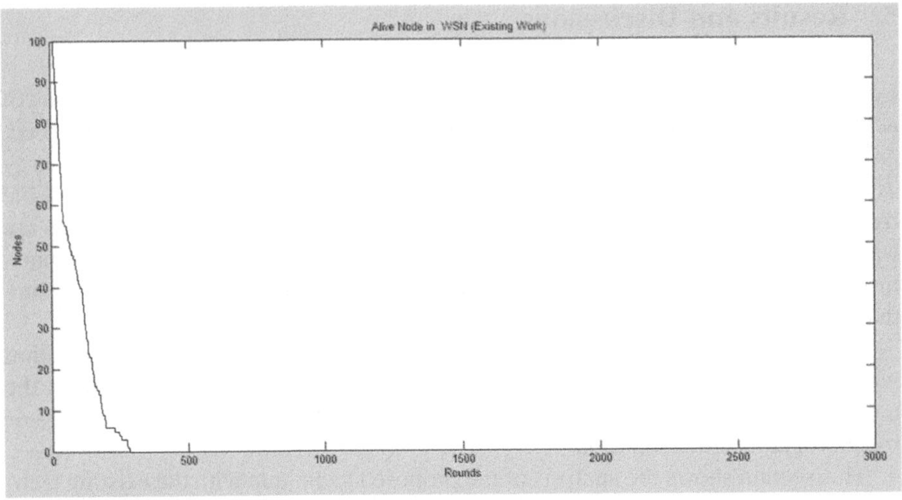

Fig. 6 Alive node analysis (existing approach)

(b) *Alive nodes*

Figure 6 shows the active nodes over the network in case of exiting approach. The graph indicates the count of active (alive) sensor nodes. As shown, the sensor nodes become dead at the earlier stage. The overall life of the network is limited to 380 rounds.

(c) *Number of packets transmitted to base station (without aggregation)*

Figure 7 shows the overall packet transmission to the base station in the network. The graph indicates overall packet transmission in existing LEACH approach. As shown, when the complete network is alive, 100 packets are transmitted in each round. But as the nodes start to die, the network communication also suffers. After about 380 rounds, the communication is stopped.

(d) *Packet Count communicated to BS (with aggregation)*

Figure 8 shows the aggregative packet transmission to the base station within the sensor network. The graph shows actual packets communicated in case of existing LEACH protocol. Initially, the packet transmission is increased very rapidly because the complete set of nodes is existent. But with the increase in the number of dead nodes, the transmission of packets becomes slower. After the network dies, the network communication also stops. This figure is showing the packet to base station on cumulative basis.

(e) *Number of packets transmitted in network*

Figure 9 shows the aggregative communication within the sensor network. The graph shows the packet transmission for cluster-based transmission and the communication with base station.

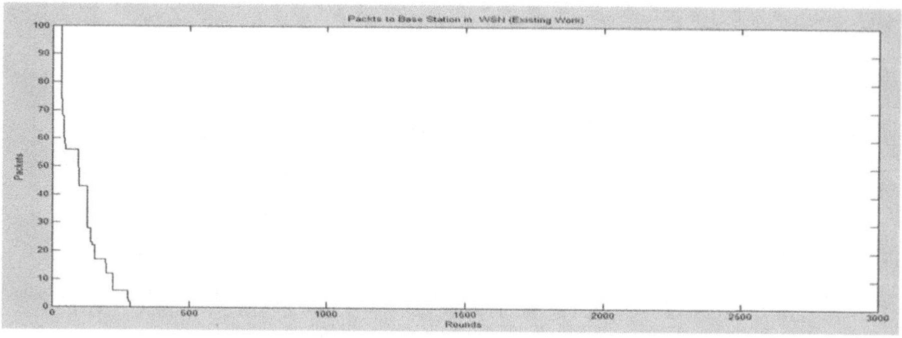

Fig. 7 Packet count communicated to BS (previous approach)

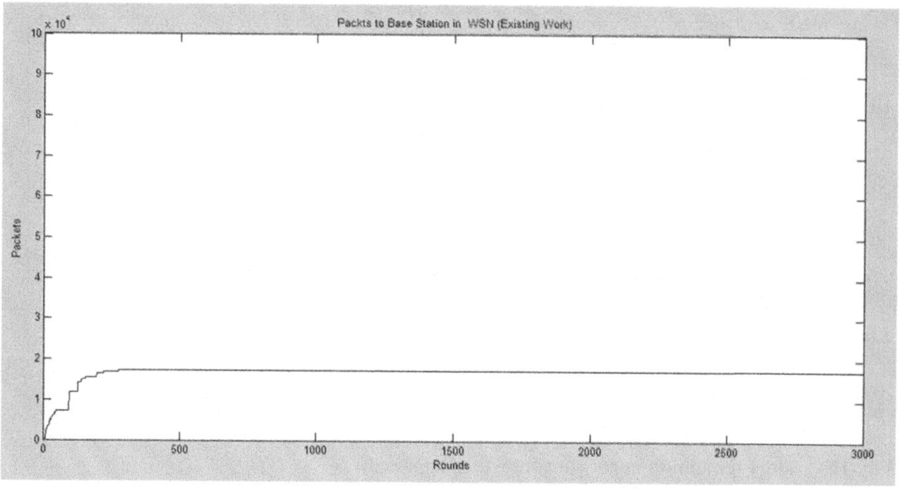

Fig. 8 Packets transmitted to base station – with aggregation (existing approach)

(f) *Energy remaining in network*

Figure 10 shows the energy remaining in the network. Each node is assigned 1 Joule of energy in the beginning, but as the communication is done, some amount of energy starts dissipating. In existing work, the complete energy is lost at 380 rounds.

In case of the currently discussed technique, results are discussed below:

(a) *Inert nodes*

Figure 11 shows the inert nodes over sensor network in case of current technique. It may be noted that the energy balance communication is performed. No node is inert up to 480 rounds, and at the end of 3000 rounds, only 38 nodes are dead.

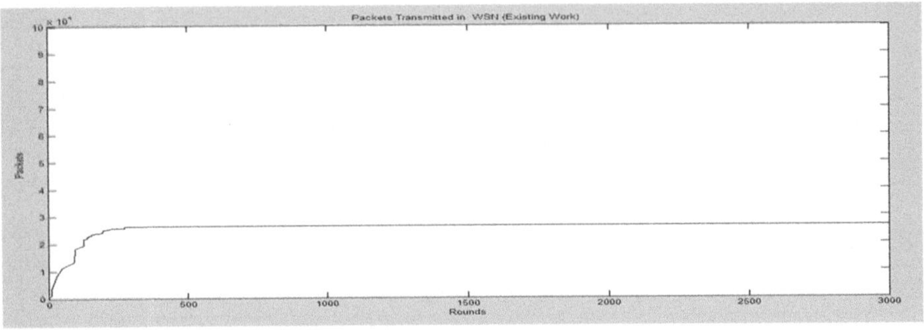

Fig. 9 Packets transmitted in network (existing approach)

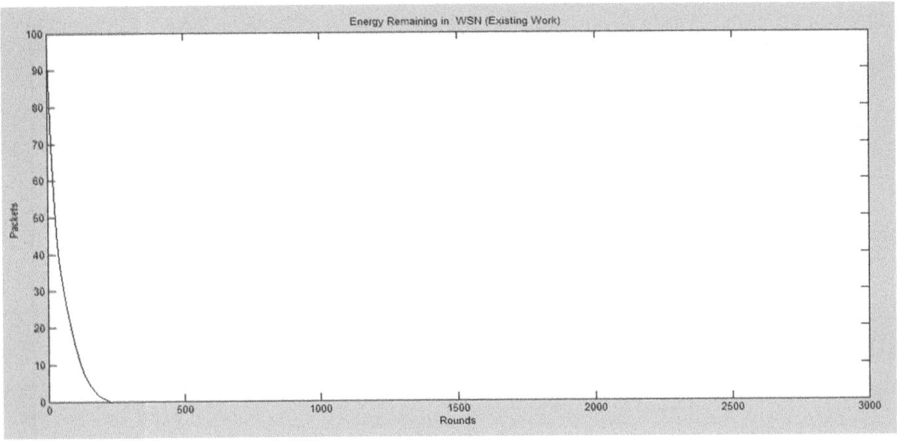

Fig. 10 Energy remaining in network (previous approach)

(b) *Alive nodes*

Figure 12 shows the alive nodes in sensor network in case of current technique. It may be seen that the network life is increased. At the end of 3000 rounds, about 62 nodes are still alive.

(c) *Packet count communicated to BS (without aggregation)*

Figure 13 shows the packet count transferred to BS in the network in case of current technique discussed. Because of the longevity of the network, about 62 nodes still remain existing at the end of 3000 rounds.

(d) *Packet count communicated to BS (with aggregation)*

Figure 14 shows the aggregative packet transmission to the base station in the current research. The packet transmission has been higher than that in the previous work.

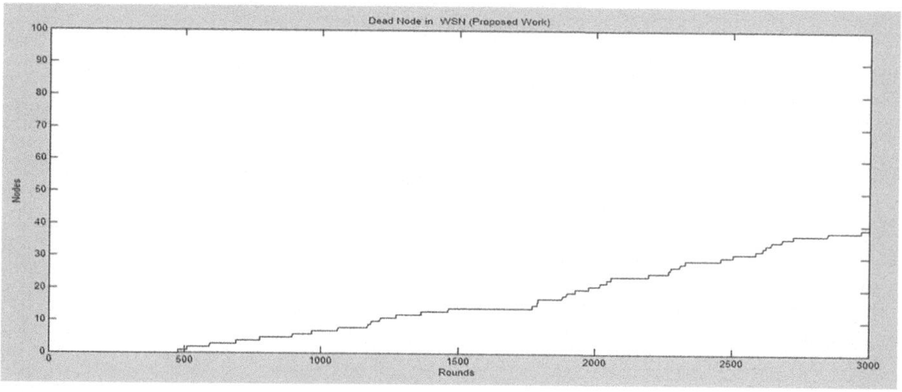

Fig. 11 Inert node analysis (proposed approach)

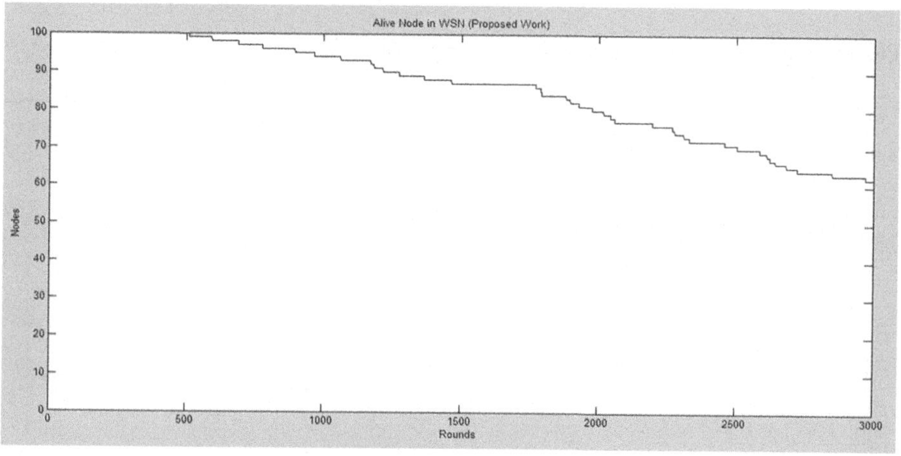

Fig. 12 Alive node analysis (proposed approach)

(e) *Number of packet transmitted in network*

Figure 15 shows the aggregative packets communicated over the network in the present research. The packet communication is increased. This packet communication includes the cluster-based transmission as well as packets transmission to the base station.

(f) *Energy remaining in network*

Figure 16 shows the energy remaining in case of proposed work. The exhaustion of energy in an individual round is very low resulting in some residual energy over the network at the end of 3000 rounds.

It may be concluded from the above graphs that:

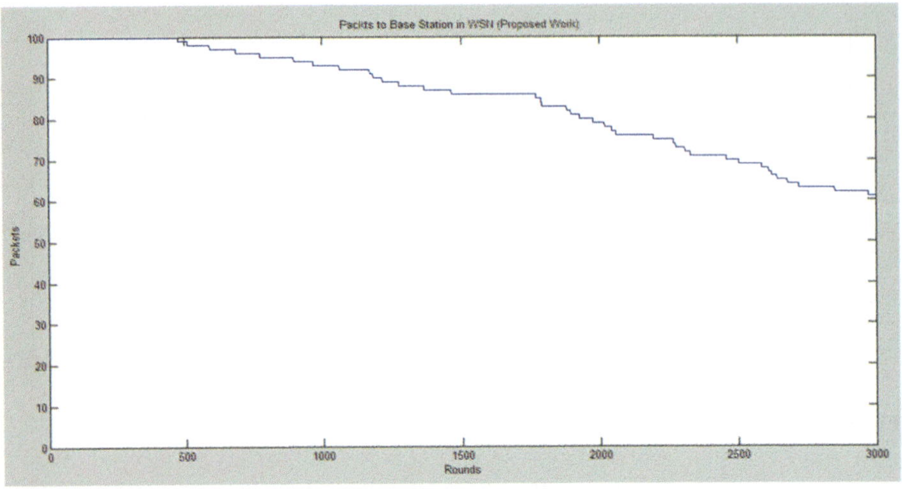

Fig. 13 Packet count communicated to BS (current technique)

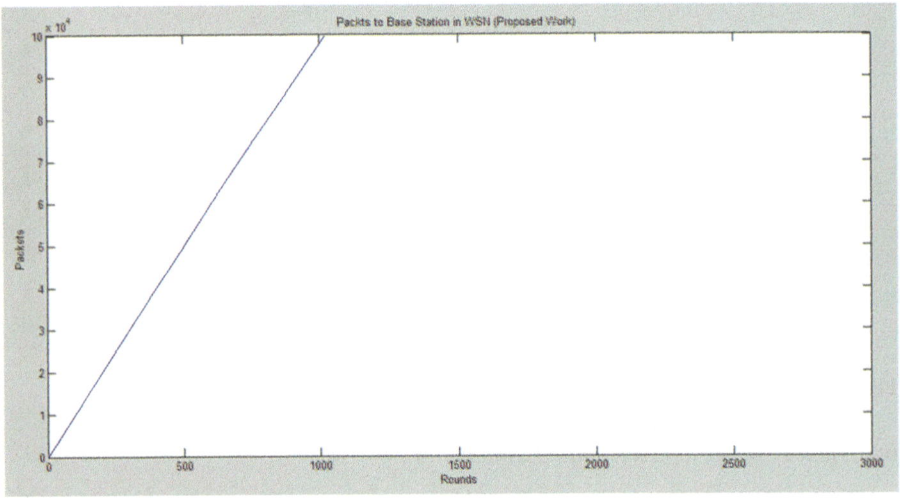

Fig. 14 Packets transmitted to base station – with aggregation (proposed approach)

(i) In the existing approach, the nodes start becoming dead in the starting rounds, while in the proposed work, the nodes start becoming dead after 480 rounds.
(ii) In the existing approach, the network stays alive up to 380 rounds. But in this proposed work, the network remains alive even after the completion of 3000 rounds.

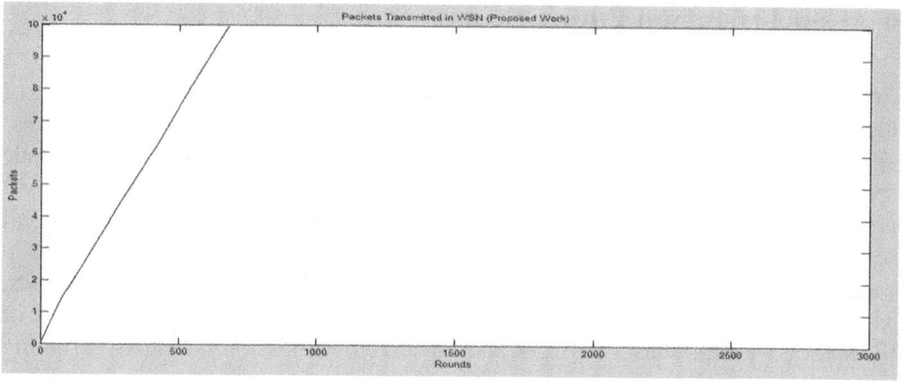

Fig. 15 Packets transmitted in WSN (proposed approach)

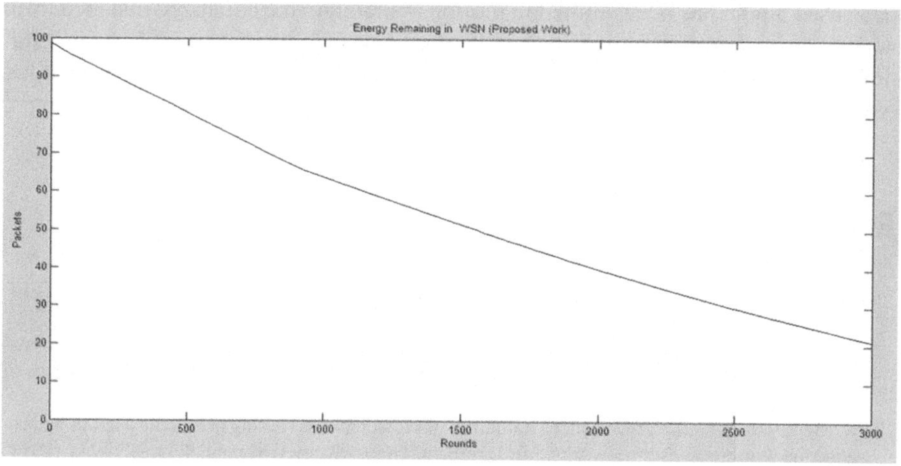

Fig. 16 Energy remaining in WSN (proposed approach)

(iii) In the existing approach, the packet transmission is performed up to 380 rounds. But in the proposed work, the packet transmission remains continued even after the completion of 3000 rounds.
(iv) In the proposed approach, the packet transmission in ad hoc wireless sensor network is much better.
(v) In the proposed approach, the energy consumption is lower than in the existing approach.

So, the proposed approach is far better than the existing approach considering the above results.

6 Conclusion and Future Scope

Energy efficiency [55, 56] is a very crucial and significant parameter in any type of network. With every communication done by the nodes, some part of energy is dissipated, and hence there is a need to save the energy. In this research, a relay node specification-based work has been implemented to achieve the same, and a hot spot area specification has been maintained to have the direct communication with all sensors and to avail the information directly to them. To avail this information to each node effectively, the relay nodes have been laid in the form of cluster heads. The obtained results justify that the current research has increased the durability of the ad hoc wireless sensor network and effective transmission for LEACH routing protocol.

The presented work of relay-based clustering approach to reduce the multi-hop communication may be extended further. In this work, a single mobile agent has been used, and it has been defined in motion to maintain the equalized distance from nodes. For instance, many agents can be used (instead of single mobile agent) stationary or in motion for the same purpose. Also, the work discussed may be extended based on the collective communication rather than based on a clustered planning.

References

1. I.F. Akyildiz, W. Su, Y. Sankarasubramaniam, E. Cayirci, A survey on sensor networks. IEEE Commun. Mag. **40**(8), 102–114 (2002)
2. R. Meelu, R. Anand, Energy efficiency of cluster-based routing protocols used in wireless sensor networks, in *AIP Conference Proceedings* (Vol. 1324, No. 1) (American Institute of Physics, 2010), pp. 109–113
3. R. Meelu, R. Anand, Performance evaluation of cluster-based routing protocols used in heterogeneous wireless sensor networks. Int. J. Inform. Theory Knowl. Manag. **4**(1), 227–231 (2011)
4. H. Qi, F. Wang, Optimal itinerary analysis for mobile agents in ad hoc wireless sensor networks, in *Proceedings of the IEEE 2001 International Conference on Communications (ICC 2001), Helsinki, Finland* (2001)
5. D. Ganesan, R. Govindan, S. Shenker, D. Estrin, Highly-resilient, energy-efficient multipath routing in wireless sensor networks. ACM SIGMOBILE Mob. Comput. Commun. Rev. **5**(4), 11–25 (2001)
6. S. Madden, M.J. Franklin, J.M. Hellerstein, W. Hong, TAG: a tiny aggregation service for ad-hoc sensor networks. ACM SIGOPS Operat. Syst. Rev. **36**(SI), 131–146 (2002)
7. S. Madden, M.J. Franklin, J.M. Hellerstein, W. Hon, The design of an acquisitional query processor for sensor networks, in *Proceedings of the 2003 ACM SIGMOD International Conference on Management of Data* (2003), pp. 491–502
8. H. Qi, S. Iyengar, K. Chakrabarty, Multiresolution data integration using mobile agents in distributed sensor networks. IEEE Trans. Syst. Man Cybern. Part C Appl. Rev. **31**(3), 383–391 (2001)
9. K.K. Paliwal, P.R.A. Israna, P. Garg, Energy efficient data collection in wireless sensor network-a survey, in *International Conference on Advanced Computing, Communication and Networks' II* (2011), pp. 824–827

10. P. Garg, R. Anand, Energy efficient data collection in wireless sensor network. Dronacharya Res. J. **3**(1), 41–45 (2011)
11. Y.C. Tseng, S.P. Kuo, H.W. Lee, C.F. Huang, Location tracking in a wireless sensor network by mobile agents and its data fusion strategies. Comput. J. **47**(4), 448–460 (2004)
12. K.N. Ross, R.D. Chaney, G.V. Cybenko, D.J. Burroughs, A.S. Willsky, Mobile agents in adaptive hierarchical bayesian networks for global awareness, in *SMC'98 Conference Proceedings. 1998 IEEE International Conference on Systems, Man, and Cybernetics (Cat. No. 98CH36218)* (Vol. 3) (1998), pp. 2207–2212
13. L. Tong, Q. Zhao, S. Adireddy, Sensor networks with mobile agents, in *IEEE Military Communications Conference, 2003. MILCOM 2003* (Vol. 1) (2003), pp. 688–693
14. M. Chen, T. Kwon, Y. Yuan, Y. Choi, V.C. Leung, Mobile agent-based directed diffusion in wireless sensor networks. EURASIP J. Adv. Sig. Process. **2007**, 1–13 (2006)
15. H. Çam, S. Özdemir, P. Nair, D. Muthuavinashiappan, H.O. Sanli, Energy-efficient secure pattern based data aggregation for wireless sensor networks. Comput. Commun. **29**(4), 446–455 (2006)
16. G. Asada, M. Dong, T.S. Lin, F. Newberg, G. Pottie, W.J. Kaiser, H.O. Marcy, Wireless integrated network sensors: low power systems on a chip, in *Proceedings of the 24th European solid-state circuits conference*, (1998), pp. 9–16
17. S. Juneja, A. Juneja, R. Anand, Healthcare 4.0-digitizing healthcare using big data for performance improvisation. J. Comput. Theor. Nanosci. **17**(9–10), 4408–4410 (2020)
18. A. Valehi, A. Razi, Maximizing energy efficiency of cognitive wireless sensor networks with constrained age of information. IEEE Transac. Cogn. Communi. Network. **3**(4), 643–654 (2017)
19. M. Al Ameen, J. Liu, K. Kwak, Security and privacy issues in wireless sensor networks for healthcare applications. J. Med. Syst. **36**(1), 93–101 (2012)
20. S. Bala, A.K. Verma, *Secure routing in wireless sensor networks* (Doctoral dissertation) (2009)
21. R. Anand, P. Chawla, Optimization of inscribed hexagonal fractal slotted microstrip antenna using modified lightning attachment procedure optimization. Int. J. Microw. Wirel. Technol. **12**(6), 519–530 (2020)
22. N. Sindhwani, M. Singh, A joint optimization based sub-band expediency scheduling technique for MIMO communication system. Wirel. Pers. Commun. **115**(3), 2437–2455 (2020)
23. R. Anand, P. Chawla, A review on the optimization techniques for bio-inspired antenna design, in *2016 3rd International Conference on Computing for Sustainable Global Development (INDIACom)*, (2016), pp. 2228–2233
24. N. Sindhwani, Performance analysis of optimal scheduling based firefly algorithm in MIMO system. Optimization **2**(12), 19–26 (2017)
25. P. Chawla, R. Anand, Micro-switch design and its optimization using pattern search algorithm for applications in reconfigurable antenna. Mod. Ant. Syst. **10** (2017)
26. S. Juneja, A. Juneja, R. Anand, Reliability modeling for embedded system environment compared to available software reliability growth models, in *2019 International Conference on Automation, Computational and Technology Management (ICACTM)*, (2019), pp. 379–382
27. K. Sohrabi, G.J. Pottie, Performance of a novel self-organization protocol for wireless ad-hoc sensor networks, in *Gateway to 21st Century Communications Village. VTC 1999-Fall. IEEE VTS 50th Vehicular Technology Conference (Cat. No. 99CH36324)*, vol. 2, (1999), pp. 1222–1226
28. R. Min, M. Bhardwaj, S.H. Cho, E. Shih, A. Sinha, A. Wang, A. Chandrakasan, Low-power wireless sensor networks, in *VLSI Design 2001. Fourteenth International Conference on VLSI Design* (2001), pp. 205–210
29. A. Manjeshwar, D.P. Agrawal, APTEEN: A hybrid protocol for efficient routing and comprehensive information retrieval in wireless sensor networks, in *Parallel and Distributed Processing Symposium, International*, vol. 3, (IEEE Computer Society, 2002), pp. 0195b–0195b

30. R. Lin, Z. Wang, Y. Sun, Energy efficient medium access control protocols for wireless sensor networks and its state-of-art, in *2004 IEEE International Symposium on Industrial Electronics*, vol. 1, (2004), pp. 669–674
31. J. Dunlop, DIAS: adaptive communications for wireless sensor networks, in *2006 Wireless Sensor Networks Conference*, (2006), pp. 1–34
32. H. Chuanhe, C. Yong, S. Wenming, Z. Hao, A trusted routing protocol for wireless mobile ad hoc networks, in *2007 IET Conference on Wireless, Mobile and Sensor Networks (CCWMSN07)*, (2007), pp. 406–409
33. S. Varma, N. Nigam, U.S. Tiwary, Base station initiated dynamic routing protocol for heterogeneous wireless sensor network using clustering, in *2008 Fourth International Conference on Wireless Communication and Sensor Networks*, (2008), pp. 1–6
34. Y.Q. Zhang, L. Wei, Improving the LEACH protocol for wireless sensor networks, in *IET International Conference on Wireless Sensor Network 2010 (IET-WSN 2010)*, (2010), pp. 355–359
35. H. Kim, J. Kim, Energy-efficient resource management in wireless sensor network, in *2011 IEEE Topical Conference on Wireless Sensors and Sensor Networks*, (2011), pp. 69–72
36. A. Berger, A. Pötsch, A. Springer, Real-time data collection in a spatially extended TDMA-based wireless sensor network, in *2012 IEEE Topical Conference on Wireless Sensors and Sensor Networks*, (2012, January), pp. 41–44
37. K.K. Gagneja, K.E. Nygard, Heuristic clustering with secured routing in heterogeneous sensor networks, in *2013 IEEE International Workshop on Security and Privacy of Mobile, Wireless, and Sensor Networks (MWSN)*, (2013), pp. 9–16
38. L. Zhang, C.L. Guo, X. Liu, C.Y. Feng, An energy efficient cooperation relay scheduling scheme in multi-hop cognitive vehicular networks, in *2014 International Conference on Wireless Communication and Sensor Network*, (2014), pp. 468–473
39. M. Ababou, R. El Kouch, M. Bellafkih, N. Ababou, Energy-efficient routing in delay-tolerant networks, in *2015 Third International Workshop on RFID and Adaptive Wireless Sensor Networks (RAWSN)*, (2015, May), pp. 1–5
40. Z. Andleeb, M.R. Anjum, M.U. Sardar, Study the impact of multiple mobile sinks on lifetime of wireless sensor networks, in *2016 Sixth International Conference on Innovative Computing Technology (INTECH)*, (2016, August), pp. 418–422
41. K. Li, W. Ni, L. Duan, M. Abolhasan, J. Niu, Wireless power transfer and data collection in wireless sensor networks. IEEE Trans. Veh. Technol. **67**(3), 2686–2697 (2017)
42. R. Ngangbam, A. Hossain, A. Shukla, An improved clustering based hierarchical protocol for extending wireless sensor network lifetime–EG LEACH, in *2018 IEEE International Conference on System, Computation, Automation and Networking (ICSCA)*, (2018, July), pp. 1–5
43. I. Banerjee, P. Madhumathy, An agent cluster based routing protocol for enhancing lifetime of wireless sensor network, in *2019 1st International Conference on Advanced Technologies in Intelligent Control, Environment, Computing & Communication Engineering (ICATIECE)*, (2019), pp. 265–268
44. Z. Zhang, R. Luo, W. Fu, Energy saving algorithm of wireless network nodes in cluster, in *2020 International Wireless Communications and Mobile Computing (IWCMC)*, (2020), pp. 394–397
45. C. Intanagonwiwat, R. Govindan, D. Estrin, J. Heidemann, F. Silva, Directed diffusion for wireless sensor networking. IEEE/ACM Trans. Networking **11**(1), 2–16 (2003)
46. M.H. Anisi, A.H. Abdullah, S. Abd Razak, Energy-efficient data collection in wireless sensor networks. Wirel. Sens. Netw. **3**(10), 329–333 (2011)
47. A.S. Poornima, B.B. Amberker, Agent based secure data collection in heterogeneous sensor networks, in *2010 Second International Conference on Machine Learning and Computing*, (2010), pp. 116–120
48. H. Chen, H. Mineno, T. Mizuno, Adaptive data aggregation scheme in clustered wireless sensor networks. Comput. Commun. **31**(15), 3579–3585 (2008)

49. M. Chen, T. Kwon, Y. Choi, Data dissemination based on mobile agent in wireless sensor networks, in *The IEEE Conference on Local Computer Networks 30th Anniversary (LCN'05)*, (2005), pp. 1–2
50. M. Chen, T. Kwon, Y. Yuan, V.C. Leung, Mobile agent based wireless sensor networks. J. Comput. **1**(1), 14–21 (2006)
51. H. Sin, J. Lee, S. Lee, S. Yoo, S. Lee, J. Lee, S. Kim, Agent-based framework for energy efficiency in wireless sensor networks. World Acad. Sci. Eng. Technol. **46**, 305–309 (2008)
52. W.H. Liao, Y. Kao, C.M. Fan, Data aggregation in wireless sensor networks using ant colony algorithm. J. Netw. Comput. Appl. **31**(4), 387–401 (2008)
53. D. Chander, R. Kumar, Energy efficient buffer Management for Group Communication in MANETs. Int. J. Adv. Res. Comput. Sci. **8**(1) (2017)
54. D. Chander, R. Kumar, Analysis of scalable and energy aware multicast routing protocols for MANETs. Indian J. Comput. Sci. Eng. (IJCSE) **8**(3) (2017)
55. A. Dahiya, R. Anand, N. Sindhwani, D. Kumar, A novel multi-band high-gain slotted fractal antenna using various substrates for X-band and Ku-band applications. MAPAN (2021). https://doi.org/10.1007/s12647-021-00508-3
56. A. Dahiya, R. Anand, N. Sindhwani, D. Deshwal, Design and construction of a low loss substrate integrated waveguide (SIW) for S band and C band applications. MAPAN. 36, 355–363 (2021). https://doi.org/10.1007/s12647-021-00449-x

Security Challenges in 5G Network

Gitimayee Sahu and Sanjay S. Pawar

1 Introduction

The 5G is the next-generation wireless technology having the following use cases, eMBB (enhanced mobile broadband), URLLC (ultrareliable low latency communication), and mMTC (massive machine type communication). eMBB assures high-speed Internet up to 10 Gbps downlink and 1 Gbps uplink speed which is 100 times higher than 4G LTE-A [1]. In LTE-A the downlink user experienced data rate is up to 100Mbps and uplink data rate up to 50 Mbps. Peak downlink spectral efficiency is 30 bps/Hz, and peak uplink spectral efficiency is up to 15 bps/Hz. The minimum requirement of user plane latency of eMBB is 4 milliseconds, user plane latency of uRLLC is up to 1 millisecond, and control plane latency is up to 20 milliseconds [2]. The control plane is used to carry signaling traffic, the user plane carries the data traffic, and the management plane carries administrative traffic [3]. The user plane, control plane, and management planes can be attacked one at a time or simultaneously which affects the entire network system. The requirement of connection density is up to one million devices per sq. km. The bandwidth of the frequency channel is at least 100 Mbps [4].

As industrial technology expands horizontally and vertically, augmented reality (AR)/virtual reality (VR), autonomous vehicles, remote surgery, smart city, smart traffic management, smart agriculture, digital locker, Internet of things (IoT) and cloud computing, industrial IOT (IIOT), and high-speed bullet trains are few examples (shown in Fig. 1) that require high-speed Internet and ubiquitous network connectivity to achieve the new momentum. The machine-to-machine (M2M)

G. Sahu (✉) · S. S. Pawar
Department of EXTC, UMIT, SNDT Women's University, Mumbai, India

M. M. Ghonge et al. (eds.), *Software Defined Networking for Ad Hoc Networks*,
EAI/Springer Innovations in Communication and Computing,
https://doi.org/10.1007/978-3-030-91149-2_4

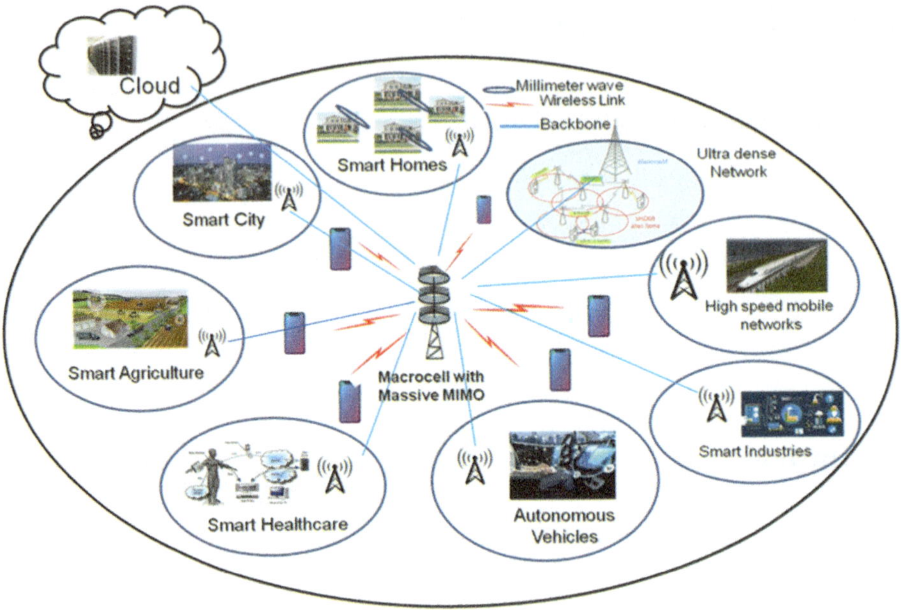

Fig. 1 5G wireless network architecture

communication gives a new dimension to the 5G technology compared to the legacy network.

5G will be demanding since it offers automation, digitization of devices, as well as smart transportation of industries. The expansion of 5G architecture, new technologies, and new business will have to face enhanced challenges of security and privacy protection. It introduces IMSI security through encryption. All the data traffic traveling through the radio network has to be encrypted and mutually authenticated from device to network. eSIM removes the requirement of altering the SIM card on the mobile handset.

Though IoT technology already exists in 3G/4G wireless networks, the number of devices and amount of data increased exponentially in 5G. The security controls may change gradually. It requires deployment, management, and secure coding during the lifetime of the network. IoT architecture is vulnerable in three common scenarios, viz., (i) remote attack on the devices through the applications running on the network via the Internet and physically also, (ii) attack on cloud or service platform, and (iii) attack on the communication link by air interface and wireless LAN (WLAN) [5].

The key elements of 5G security include authentication and key agreement (AKA), data integrity, cryptographic algorithm, and enhanced security for connectivity to other networks. It also includes home network control, and detection of invalid base stations depends on UE data and subscription concealed identifier (SUCI). Authentication and key management (AKM) is the primary parameter of mobile networks to safeguard the user, network, and information exchange.

5G has improved consolidated authentication procedure compared to 4G. It has superior UE identity protection, sophisticated home network control, high key separation in different key derivations. In addition to AKA, non-AKA methods such as EAP-TLS (extensible authentication procedure and transport layer security) can also be used. The old devices not having USIM behind the femtocell, i.e., Home eNodeB (HeNB) is directly connected to the 5G packet core using the EAP authentication process. The data integrity method in 5G provides an improved cryptographic algorithm with 256-bit key security against data tampering at the air interface. The cryptographic algorithm is used in 5G for encryption and decryption at the air interface. Ericsson uses a new rapid cipher "SNPW-V" for 5G mobile systems.

Horizontal security in 5G enhances privacy during data transfer to other networks. The transferred data are highly confidential and protected with integrity. The transport layer plays a significant role in transferring data from one network to another horizontally. A similar kind of authentication has to be followed during switching from one cell to another or when the user enters from one network to the other network during handover.

The security of 3GPP during the reception of RF information by the user devices is an integral part of the cellular network of all generations. One of the security concerns is the fingerprint of false base stations. To detect the false base station, IMSI (international mobile subscriber identity) catcher methods using subscriber long-term identifier (LTI) and frequent variation of short-term identifier (5G-STI) are used. These two methods enhance user protection against the false base station [6, 7]. The data set for 5G security architecture and procedure is implemented regarding 3GPP TS 33.501. It refers to SUCI calculation of 5G UE for IMSI-based SUPI (subscription permanent identifier) and Elliptic Curve Integrated Encryption Scheme (ECIES) Profile A [8].

The enabling technologies of 5G include software-defined networking (SDN), network function virtualization (NFV), network slicing, mobile edge computing (MEC), and control plane and user plane separation (CUPS). Network slicing mechanism slices the available resources, i.e., Internet, mobile, and cloud architecture for various private networks. New protocols were defined for the control, user, and management plane. The core network is associated with cloud services for reduced latency. Extensible authentication protocol (EAP), authentication, and key agreement (AKA) methods are used to support public key infrastructure (PKI)-based authentication [9, 10].

Furthermore, the network slicing technique invokes the mobile network operator (MNO) to slice the network as per the priority of different use cases and requirements. McQuire from Microsoft addresses that network slicing enhances security to the network since it uses its private channel [11]. For remote surgery, the network slice should account for mutual identification and authorization to prevent a "Man-in-the-middle (MITM)" attack, but the AR/VR application may not require a similar kind of security.

5G security is a shared responsibility between the network operator, architecture, and infrastructure.

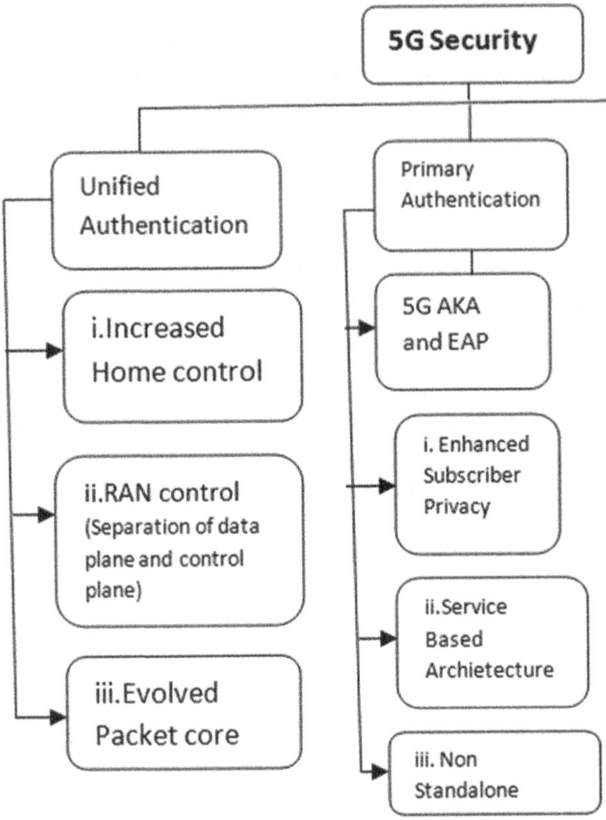

Fig. 2 5G security procedure

The 5G security consists of unified authentication, primary authentication, and secondary authentication as shown in Fig. 2. Unified authentication includes increased home control and RAN control which separates the data plane and control plane. Primary authentication consists of a 5G authentication key agreement (AKA) and extensible authentication protocol (EAP). Secondary authentication includes visibility and configurability, initial NAS message, and PLMN interconnect security. The subscription permanent identifier (SUPI) is of two kinds, IMSI and Network Access Identifier (NAI). IMSI includes MCC, MNC, and mobile subscriber identifier (MSIN). In non-standalone mode, the 4G eNodeB and the 5G gNodeB were connected to the 4G core network.

The 5G network architecture is ultradense, heterogeneous, and multitier that supports massive MIMO with beamforming, non-orthogonal multiple access (NOMA), and full-duplex communication. The physical layer includes user equipment (mobile devices, IoT devices, and various sensors of industries) which are connected to the radio access network (gNodeB) through an air interface as shown in Fig. 3. The gNodeB is interfaced with RAN security. The network can be sliced for different

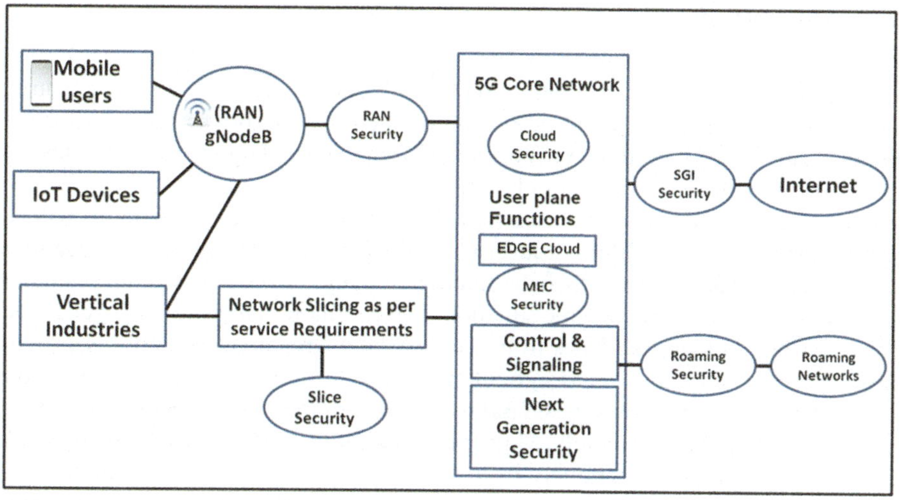

Fig. 3 Location of security points at different interfaces of 5G network architecture

services for enhancing security. The core network consists of the user plane, control, and signaling plane with cloud and mobile edge computing (MEC) security. It integrates solutions group international (SGI), security with Internet cloud, and roaming networks. The SGI security is mainly used for financial transactions in banks and account-related services. The function can be performed seamlessly and transparently to manage the entire organization with integrity.

The security in 5G network is multilayered and multitier as well. Along with application security, we have to consider security at cloud, various services, and device terminals. Security at core network and between different networks needs to be contemplated. The radio interface security should be tightened between the network and device, i.e., device to device (D2D) at air interface level. The process of security includes encryption, authentication, and key agreement (AKA). Security in different layers includes (a) big data security, terminal, and cloud security at higher layers; (b) routing security, core network, and internetwork security at middle layers; and (c) physical layer security at the lower layer. The 5G network has different use cases such as smart city, smart healthcare, smart traffic management, smart agriculture, smart industries, and military and defense applications where security will be a main challenge. To meet the challenges, the network security at evolved packet core (EPC) and multitier security architecture are the imperative elucidation for it.

This paper is organized in the following manner. Section 1 presents introduction, Sect. 2 presents motivations, Sect. 3 presents objective, Sect. 4 presents literature review, Sect. 5 presents system model, Sect. 6 presents results and discussion, and Sect. 7 presents conclusion and future scope.

2 Motivations

As 5G technology is service centric, 4G is user-centric, and 3G is operator centric; hence in 5G, emphasis should be on safety, security, and privacy of data for the enabled services. Much tighter security is indeed expected compared to the legacy network. 5G will serve vertical industries; hence various data-centric (using artificial intelligence (AI), machine learning (ML) and data mining) applications will be created which requires different level of security. High-speed mobile applications need more stringent security than the Internet of things (IoT). IoT is getting impulse now in this COVID-19 pandemic situation. The enormous amount of data generated from sensors on real-time basis from wireless body area network (WBAN), smart hospital, smart city, and smart agriculture and industries has to be stored in IoT cloud. The devices and sensors should be attached to the cloud network and can be accessed and operated remotely. Many people want to connect; access the network and can give instruction remotely from smart home. Hence significant authentication procedures for enhanced security are used such as face identification and biometric techniques, i.e., retinal and finger print scan, one time password (OTP), and two-step verification procedures are some of the methods. This increases security and reduces the unauthorized access of the IoT devices.

3 Objective

The objective of the research work is to list out different security threats in the 5G network and related challenges and how to overcome these challenges are discussed. 5G technology is a paradigm shift in cellular communication which connects device to device, device to human, and human to human, i.e., Internet of everything (IoE). Hence security becomes a vulnerable issue to maintain the privacy of the user and data sensed by the devices. Thus we need to figure out what are the different methodologies need to be used for increasing security. In this research work, the use case of security challenges during handover by a user from one cell to another cell is explained with proper experimentation procedure.

4 Literature Review

Mobile communication system has emerged from innovations of wireless technology from 2G, 3G, and 4G to maintain the pace with exponentially rising data and voice over LTE (VoLTE) traffic as shown in Fig. 4. In 2G, the authentication procedure is one way, but in 3G and 4G, mutual or two-way authentication process exists with key separation and subscriber identity protection. When the user moves vertically downward from 5G to 4G/3G network, there should be seamless handover

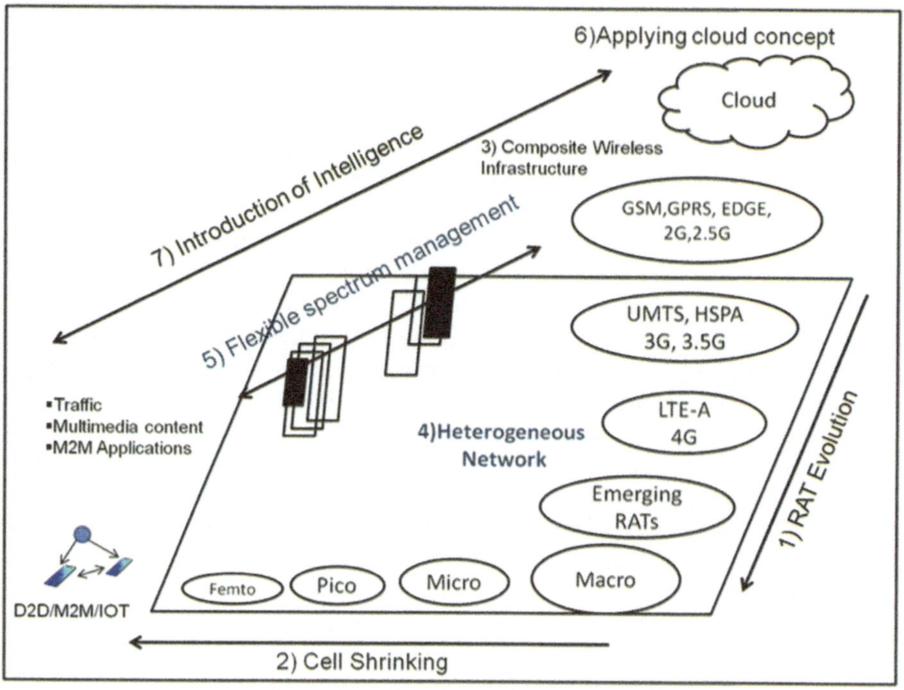

Fig. 4 Evolution of wireless technology from 2G to 5G

with reduced data rate. 5G technology is leading paradigm shift of the existing LTE-A wireless technology. 5G new radio (NR) supports millimeter wave (mmWave), massive MIMO, and advanced spectrum sharing mechanism. It should provide enhanced user protection than the current cellular network. It will offer unprecedented opportunities for enhancing the security at network and service levels. Following are some of the literature reviews related to 5G security and related challenges.

The 5G security authentication procedure and the regulatory impact plays a vital role for the privacy and security of 5G network [12, 13].

Johnson J. et al. in [14] explain millimeter wave phased array to achieve 5G mobile network's vision of increasing capacity, high data rate, and low latency. It increases the security at the physical layer arising from millimeter wave communication at the air interface.

Xiaodong et al. in [15] explain SDN-based security architecture for 5G. The deployment of software-defined networking (SDN), network function virtualization (NFV), IoT and cloud computing enables security and privacy issues from the shared environments and advanced players with objectives on privacy. The solutions are SDN-based privacy-aware routing mechanisms, hybrid cloud approach, privacy by design, and software-defined privacy. The total solution involves regulation and legal framework. The centralized security server interacts with SDN controller. The dynamic network security services are provided to the mobile user on-demand basis. Amir et al. in [16] explain millimeter wave phased arrays for physical layer which facilitate high data rate and low latency with security. In this research work, issues related to capacity and privacy of 5G network are discussed and propose a solution to enhance the security by deploying phased array antenna.

Jong ho et al. in [17] explain security policies in 5G smart city using block chain technique safety measures designed for user authentication to provide reliability, transparency, and efficiency for the digital urban environment.

Jin cao et al. in [18] explain various security aspects of 5G wireless network. 5G supports different applications such as massive Internet of things (IoT), D2D communication, AR/VR, machine-to-machine (m-to-m) communication, viz., driverless cars and remote surgery. The slicing of network resources can be done for different services [19]. High security is required in 5G network for different mission critical applications.

Garima et al. in [20] explain security issues in the heterogeneous ultradense network. As the user density along with network density increases, it becomes important to enhance the security level identifying the sensitive zones and heightening the handover points for high-speed users.

Ghada et al. in [21] explain 5G securities in reference to 3G and 4G architecture. Ning W. et al. in [22] explain physical layer security to fulfill the increasing demands of Internet of things (IOT). The characteristics of 5G IoT are reviewed under some applications. The security threats from 5G IoT physical layer is introduced and categorized as per different functions.

Fuwen Liu et al. in [23] enable users to access the network which is a primary requirement. There are many security challenges faced by the user in 3G and 4G networks, such as, leakage of long-term key, privacy of subscriber identifier, insecurity link of mobile network operators and likability attacks. The 5G-enhanced securities mitigate all issues using in one scheme without relying on public key infrastructure (PKI). Asim et al. in [24] explain the key management is the primary issue. A unique secret key can be shared between two communicating parties in the air interface. Security is the key exchange done by preequalization, guaranteed communication by the use of low-density parity check (LDPC) codes.

Yongpeng et al. in [25] discusses the physical layer security (PLS) considering the randomness of the transmission channel in air interface. The physical layer security includes technologies such as, full-duplex, massive MIMO, non-orthogonal multiple access (NOMA), heterogeneous network (HetNet), millimeter wave technology, etc. Li Sun et al. in [26] explain ultra-high user density and exponentially rising data traffic make the network security significantly high. Encryption methods are used for higher layer security. The PLS solution prevents reduction of the signal quality in the air interface by the malicious users. It avoids the distribution and management of secret key and offers flexible security by adaptive transmission protocol design. The application of PLS to 5G provides guaranteed solution for the security threats. Morteza S. in [23] explains different spectrum sharing cognitive radio methods. In the first case, the high-priority devices share the spectrum with the low-priority devices. In the other case, the high-priority device needs to decode the message of the low-priority system. The low-priority device assists the high-priority devices to get improved security. Qi Fang et al. in [24] propose unified security architecture which integrates unified and additional authentication, encryption, access, and integrated security.

5G supports smart city, smart home, wireless body area network (WBAN), smart hospital, digital twin technology, and industrial IoT (IIoT) where all the data were put in cloud for remote monitoring and controlling on real-time basis; hence information and data security becomes a vulnerable issue.

4.1 Security Threats of 5G Heterogeneous Network

5G has designed for security controls of consigning many vulnerability issues. These controlling methods are (i) new authentication capabilities and (ii) improved subscriber identity protection with supplementary security system. 5G technology guarantees the mobile communication network with remarkable opportunity to exhilarate the enhanced security levels. It offers defending measures to the mobile industry to restrict the effect of threats. The next-generation technology introduces new threats for the cellular industry. 5G standard develops adoptive security design principle which leads to mutual authentication. It establishes end-to-end secured relationship between the sender and receiver. The encryption and decryption methodology in inter- and intra-tier network ensures the ciphered information is worthless when ambushed.

Figure 5 shows the 5G security architecture which includes (i) network access security, (ii) network domain security, (iii) user domain security, and (iv)

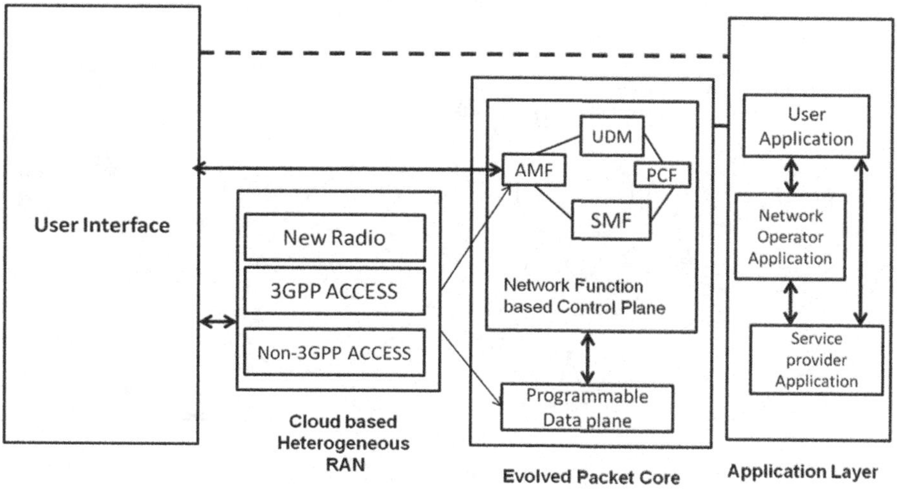

Fig. 5 5G security architecture

application security. Network access security is between the user equipment and the core network to prevent attack on the radio access link. The access security emphasizes on confidential and integrity protection between the UE and the cloud RAN, user identity and location confidentiality, authentication of various entities, and user and signaling data confidentiality. Network domain security counts the set of security features which protects against the attack between different nodes during signaling information exchange. This ensures security between cloud RAN and evolved packet core (EPC). The security in user domain is the authentication and authorization process between the user and the evolved packet core (EPC) network. Authentication between user and the network operator and the service provider is also essential for availing different applications with security. The new identity management method enhances the security of 5G network. Application domain security is the group of authentication message exchanges between the user and application server provided by the network operator.

There are five different attacks can happen in 5G wireless network as shown in Fig. 6:

(i) Eavesdropping.
(ii) Jamming.
(iii) DDOS.
(iv) Man-in-the-middle attack (MITM).
(v) IP spoofing.

Eavesdropping is the method where the attacker steals other secured information without their consent. Jamming can agitate the communication between the users.

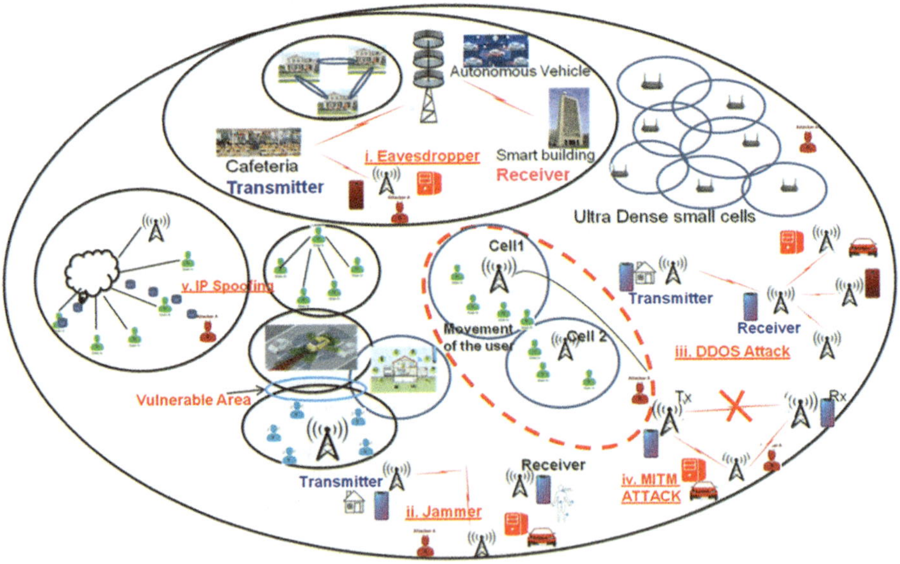

Fig. 6 The different attacks in 5G heterogeneous network

The vicious node will create interference and reduce the signal strength and disturb the communication between genuine users. It may avoid the authorized users to access the resource blocks. Direct sequence spread spectrum (DSSS) and frequency-hopped spread spectrum (FHSS) widely used for secure communication and prevents jamming. Denial of service (DOS) and distributed denial of service (DDOS) attacks can happen at different layers which causes serious effects to the entire wireless network. Since 5G network is ultradense, hence massive amounts of sensors will be interconnected among each other; hence DDOS attack creates vulnerable issue in the network. In the MITM attack, the intruder takes over the communication channel and can able to modify, remove, and add new contents in the conversation between the genuine users. This attack is active and affects different layers of the network. The Man In The Middle (MITM) attack using false base station makes the legitimate user to associate with the false base station, by using mutual authentication and authorization process between the mobile device and base station (BS) which is used to inhibit the MITM due to false base station. IP spoofing is the conception of IP packets from a different source address in order to conceal the identity of the sender. It's a process of using false address to conjure the targeted device of the encompassing infrastructure.

5 System Model

The system model consists of three-tier HetNet having macrocell, picocell and open access femtocell. The density of the picocell λ_p is ten times higher than the density of macrocell λ_m. The density of femtocell λ_f is ten times higher than that of the picocell λ_p. The small cells are uniformly distributed in a random manner across the defined area. The simulation is carried out using Poisson point process (PPP) and Voronoi tessellation in MATLAB 14b as shown in Fig. 7.

The signal to interference plus noise ratio (SINR) of a femto user (FUE) associated with femtocell is given by.

$$\Gamma_{f,n} = \frac{p_f g_{f,n}}{\sum_{f'=1, f'\neq f}^{N} p_{f'} g_{f',\{f,n\}} + \sum_{p=1}^{m} p_p g_{p\{f,m\}} + p_t g_{t\{f,n\}} + \sigma_{f,n}^2} \tag{1}$$

where P_f is the transmitting power of the femto base station (FBS), P_p is the power of the pico base station (PBS), and P_t is the power of the macro base station (MBS). The numerator denotes the signal received by the FUE with the associated FBS. The denominator consists of the net interference to the FUE by the neighbor FBS, PBS, and the MBS as well. The noise is the additive white Gaussian noise (AWGN) $n_i \sim N(0, \sigma^2)$ having zero mean and variance σ^2 in milliwatts (mW). The path loss varies for different scenarios such as, urban, semi-urban, and rural areas.

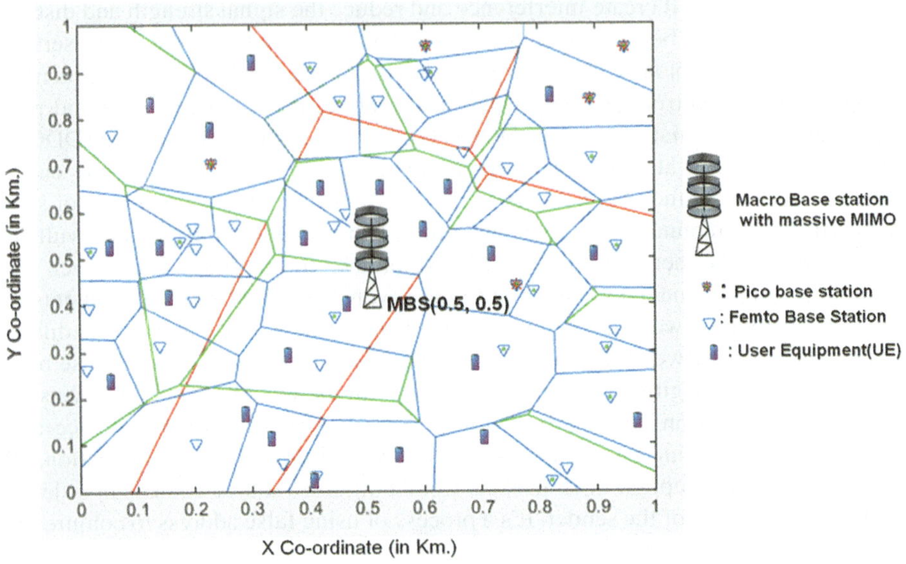

Fig. 7 Block diagram of three-tier HetNet including macrocell (red regions), picocell (green regions), and femtocell (blue regions)

The downlink capacity of the FUE served by FBS can be expressed using the Shanon's capacity formula as.

$$C_f = B_f \log_2\left(1+\Gamma_{t,f}\right) \tag{2}$$

The sum capacity of all "N" SUEs associated with FBS is given by.

$$C^f = B_f \sum_{f=1}^{N} \log_2\left(1+\Gamma_{f,n}\right) \tag{3}$$

Based on the measurement of channel quality indicator (CQI) of the radio frequency channels, the signal strength can be quantified. The estimation of channel quality is scaled between 0 and 30 dB. CQI:0 dB indicates channel quality is worst due to high interference and large distance between the user and base station. CQI:30 dB indicates best channel quality with high SINR. As per the definition by 3GPP standard, the CQI can be expressed as.

$$CQI = \begin{cases} 0; SINR \le 16dB \\ \dfrac{SINR}{1.02} + 16.62; \left(-16dB < SINR < 14dB\right) \\ 30; SINR \ge 14dB \end{cases} \tag{4}$$

as the 5G network is ultradense and there is deployment of multitier network. To obtain better signal strength and higher data rate, the distance between the access point (AP) and the user should less which shrinks the cell size. Depending on the SINR measured by the UE, handover may take place frequently from one AP to the other. The different kinds of services can be prioritized, as real-time voice calls, video calls, video conferencing, and live telecasting of events are high-priority; interactive applications such as gaming, streaming video files, email, and chatting are medium priority; downloading of movies and video files which consume lots of bandwidth is considered as background services.

As the signal strength of the user reduces with reference to the associated cell, this may happen due to high interference from the neighboring cells. Hence the user can be biased to handover to the adjacent cell:

$$CellID_i = \arg\max_k \left(RSRP_k + \alpha_k\right) \tag{5}$$

Depending on the SINR of the user, the capacity can be calculated using Shanon's law as.

$$C_i = B_i * \log_2\left(1 + SINR_i\right) \tag{6}$$

As shown in Fig. 6, the MBS is located at the center of the area, i.e., (0.5, 0.5). In the three-tier network, if the user is moving, then the velocity can be calculated by the distance covered per amount of time. The user will get handed over from one cell to the other during handover. The maximum possible attack occurs at the boundary of cell1, during handover, and at the boundary region of cell 2. These positions decided during movement of the user on the call. These regions known as perceptive areas or highly sensitive zones for the intruder to attack. The signal strength decreases at the boundary which depends on the distance between the user and the base station. Let the user is present at (x2, y2) = (0.7, 0.8) km. from the MBS which is situated at (x1, y1) = (0.5, 0.5) km. The distance can be calculated as.

$$D = \sqrt{\left(x_2 - x_1\right)^2 + \left(y_2 - y_1\right)^2} = 0.3606 Km. \tag{7}$$

The simulated scenario is of 1 Km^2, and the FBS has radius of 10 m^2; hence there will be 100 numbers of FBS in the deployed area. The PBS has radius up to 100 m^2. Hence 10 PBS will be deployed in the area. The number of handovers can be calculated as.

$$HO(\%) = \frac{HO_{FBS}}{HO_{MBS} + HO_{PBS} + HO_{FBS}} \tag{8}$$

For FBS, the percentage of handover is more, i.e., 90%; for PBS, the percentage of handover is 9%; and for macrocell the percentage is 0.9%, nearly 1%.

The algorithm describes the architecture of the network, deployment of cells as per user distribution, channel quality measurement, and SINR calculation. The attack by the intruder while handover from one cell to other may occur during the session initiation time (SIT):

Step 1: To find out the spoofing attack in the multitier heterogeneous network (HetNet).
Step 2: To calculate the percentage of handover in the multitier HetNet.
Step 3: To measure the interference and its effect in multitier HetNet.

5G technology operates in the sub 6GHz band which has maximum cell size up to 1 Km. square. However, for 28GHz to 32GHz, the maximum allowable cell size is up to 500 m. In multitier network the probability of interference increases with increase in handover.

Algorithm:

1. Beginning of the process.
2. Enter number of FBS.
3. Input the transmitting power of the FBS (P_f).
4. Input the transmitting power of PBS (P_p) and MBS (P_t).
5. Set Threshold 1 for the channel condition as Th1; set Threshold 2 for SINR as Th2.
6. Calculate the distance of the user with respect to the associated cell.
7. Measure the channel condition of the user with respect to the associated cell.
8. Calculate SINR of the user with respect to the associated cell and list out the interference from the neighboring cell.
9. If the measured channel condition is less than Th1 and SINR < Th2.
10. Then handover is initiated to the neighboring cell.
11. The session initiation time (SIT) is started and measured.
12. Whether time is synchronous with the attacker or not needs to be monitored.
13. If yes, vulnerable to attack by the intruder.
14. If no, then wait for the next SIT (handover time).
15. End of the Process

For fast-moving users the handover is happening frequently. During this process of handover, the user may get prone to security attack. The possible regions of security threat are in the overlapping area of two adjacent cells and the cell edge areas. Such regions need to be identified and marked as susceptive points. Due to location sharing application, vulnerabilities may happen by catching the current location of the user. Hence emphasis should be given to protect the physical layer air interface. The 5G network is heterogeneous and ultradense (UDN) in architecture. Thus there are various possibilities of security attack in UDN due to cell overlapping areas.

There is possibility of IP spoofing attack during data transfer as the devices are connected to the Internet cloud through IP address. So there arises a challenge to trace the attacker. The other issue is the frequent connection losses due to mobility of the user in the small cell architecture. For high-speed user the QoS degrades due

Table 1 Simulation parameters

Sl no.	Name of the parameter	Value of the parameter
1	MBS radius	500 m
2	PBS radius	100 m
3	FBS radius	10 m
4	MBS transmit power P_t	43 dBm
5	PBS transmit power P_p	23 dBm
6	FBS transmit power P_f	12 dBm
7	Path loss between MBS and MUE (urban scenario)	$15.3 + 37.6\log_{10}(R)$ R: distance between MBS and MUE
8	Path loss between MBS and FUE	$15.3 + 37.6\log_{10}(R)$, $Pl_{hw} = 10$
9	Path loss between FBS and FUE (in serving femtocells)	$15.3 + 37.6\log_{10}(R) + 0.7R$
10	Path loss between FBS and FUE (in other femtocells)	$15.3 + 37.6\log_{10}(R) + Pl_{hw}$, $Pl_{hw} = 20$
11	FBS to MUE	$15.3 + 37.6\log_{10}(R) + Pl_{hw}$ $Pl_{hw} = 10$
12	Carrier frequency (f_C)	2.4 GHz
13	Channel bandwidth (B)	10 MHz
14	Handover margin	3 dB
15	Noise power density(N)	−174 dBm/Hz
16	Number of users	50
17	Speed of the mobile user	60 Km/h

to frequent handover and association of PBS. There will be service interruption due to more signaling overhead. Session initiation protocol (SIP) is used to establish and terminate the VoLTE call session. SIP is similar to handover in wireless communication. The packet transmission using SIP can be protected using IP security (IPsec) and transport layer security (TLS) mechanism. But the disadvantage of the mobile user has to establish the tunnel again during the process of switching of the network while moving and requires more bandwidth and control channel overhead (Table 1).

6 Results and Discussion

Consider the simulation of the system model as shown in Fig. 7 where the users are distributed uniformly following the Poisson point process (PPP). As the user is moving from one cell (1) to the other cell (2), the measured SINR by the UE decreases exponentially with respect to the associated cell (1) and increases with respect to the cell (2) as shown in Fig. 8. Thus, we have to find out the exact handover point where the frequency channel detached from cell (1) and attached with cell (2) shown in Fig. 9. The accurate handover happens at 8 seconds as shown in Fig. 9 decided by the measured signal strength of the UE from both the base stations. At

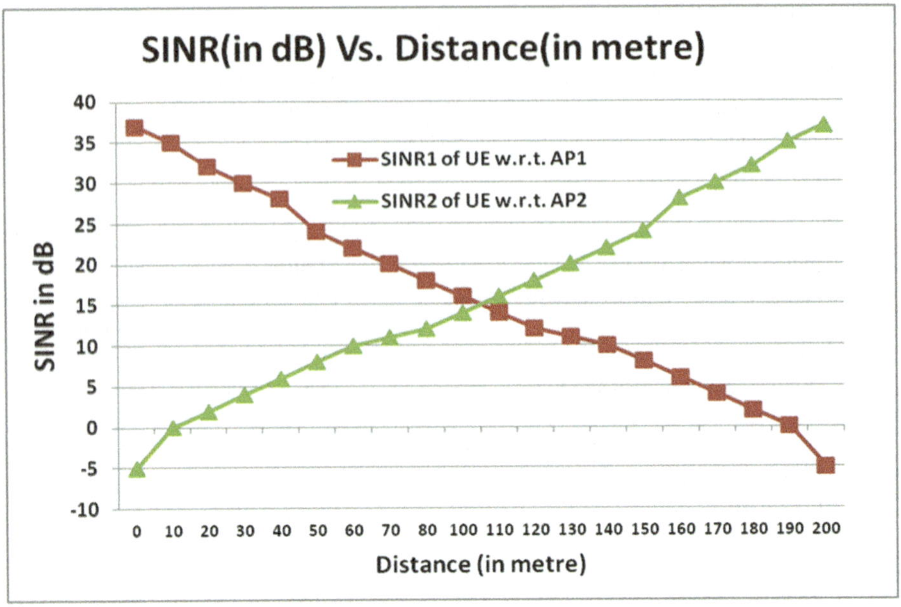

Fig. 8 Deviation of SINR of user with respect to distance

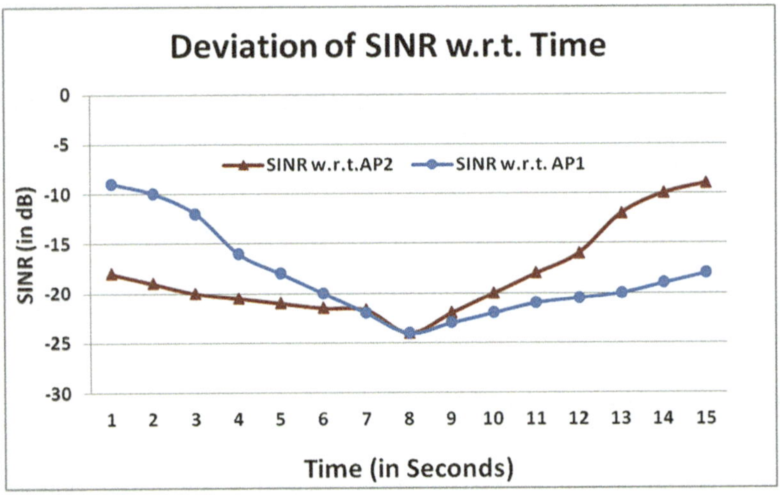

Fig. 9 Deviation of SINR with respect to time

soft handover the frequency channel will be established by the AP (2) before being detached from AP (1). Here the session initiation time (SIT) can be found out from the distance at which handover happened from the mobility scenario. In the multi-tier network, the variation of SINR measured by the UE happens due to random movement of the UE. As the user is moving further toward cell (2), the signal

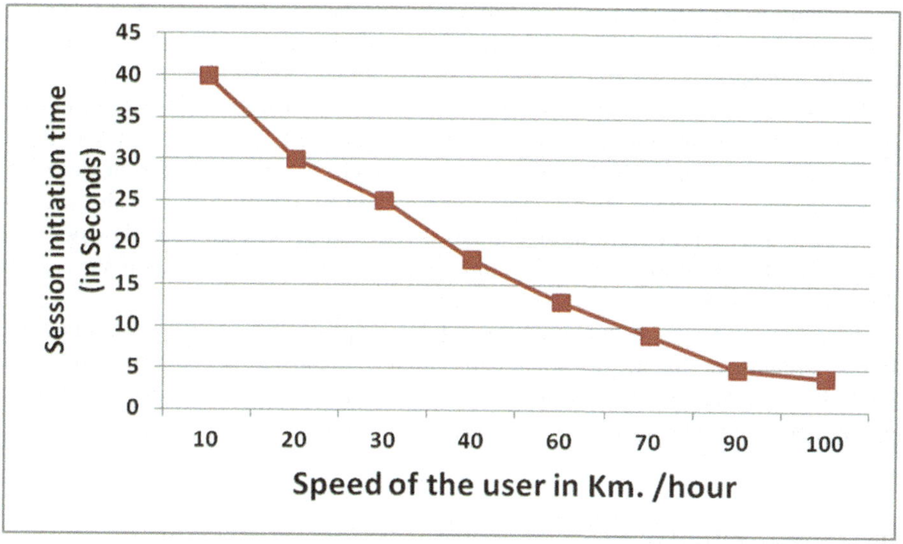

Fig. 10 Time period of spoofing attack of the mobile user

strength increases (Fig. 8). So the intruder keeps on tracking the correct point at which it can initiate spoofing of the frequency channel to get the information of the user as shown in Fig. 10. When the intruder initiates the spoofing, the CQI of the mobile user will decrease abruptly with distance.

In 5G generalized frequency division multiplexing (GFDM) modulation, unified filtered multi-carrier (UFMC), filter bank multi-carrier (FBMC), and non-orthogonal multiple access (NOMA) are used at the air interface to enhance the data rate, adequate resource allocation, and higher capacity. Low-density parity checking (LDPC) and turbo coding techniques are used for encryption and decryption at the air interface for increasing the security.

7 Conclusion

The 5G network is ultradense in architecture; due to emerging application services, strong security becomes essential for the next-generation cellular system. The upcoming 5G technology is application-centric and user-centric and requires enhancement in the security procedure to prevent the users from attacks by the intruder. Thus, 5G network needs tough authentication procedure compared to legacy network for usage of services in various sectors. The security procedure in 5G makes information and communication safe for the data hungry users from eavesdropping and spoofing attacks. Security vulnerabilities in the network rise during the process of inter-tier and intra-tier handover which occurs frequently as cell size in the network architecture reduces. There is vulnerability in attack when the session initiation time (SIT) is synchronized with the attacker's time. Thus, consistent monitoring of the signal strength in terms of SINR is essential to check the possible attack.

In this research work, the spoofing attack by the invader is discussed during handover of the mobile user from one cell to another cell. The trespasser activity is monitored continuously by measuring the channel quality and SINR of the user. When there is an attack, the signal strength reduces drastically, and the user is attached to a false base station. This attack is reduced by refreshing the key periodically and integrity protection of signaling data which is prevented as discussed in this research work.

The solution to this security threat in UDN is proposed, in order to reduce the frequent handover; the nearby small cells can form a cluster in a coordinated multipoint (CoMP) manner so that when the user is moving in the same cluster, handover will not be performed.

8 Future Challenges

The security challenges in the 5G network are related to the access network, backhaul network, and core network. The access network prone to threats related to the Internet which is intended for different types of access points, e.g., cloud RAN, picocell, microcell, and femtocell. 5G is leveraged with NFV, SDN, and cloud RAN to execute different services related to network locations. The security in the backhaul network is concerned with both radio and evolved packet core networks. The traffic toward the internal and external and the eNodeB sends the traffic to the serving gateway through GPRS tunneling protocol (GTP). The serving gateway transmits all the traffic to the public data network (PDN) gateway which then communicates to the Internet cloud. Hence the network security at the backhaul network should be enhanced through Internet protocol security (IPsec) based on GTP tunnels between the eNodeBs through the X2 interface and eNodeB and MME through the S1 interface. The EPC consists of different network entities like MME, PDN gateway, serving gateway, and home subscriber server (HSS). The IP-based core network ensures end-to-end service delivery with guaranteed QoS and heightened security of subscriber information. The 5G network is more flexible and dynamic compared to the legacy 4G network. With the diversified applications and service capabilities, the 5G network infrastructure should be more robust against vulnerable security attacks.

References

1. https://www.ericsson.com/en/mobility-report/articles/uplink-speed-and-slow-time-to-content
2. https://www.ericsson.com/en/security/a-guide-to-5g-network-security

3. G. Sahu, S.S. Pawar, Enhancing cost efficiency of femto cell by mobile traffic offloading, in *2019 IEEE 16th India Council International Conference (INDICON)*. https://doi.org/10.1109/INDICON47234.2019.9030275
4. https://5gobservatory.eu/info-deployments/5g-performance/
5. M. Liyanage, I. Ahmad, A.B. Abro, A. Gurtov, M. Ylianttila, *5G WLAN Security*, (Wiley Telecom), pp. 143–163. https://doi.org/10.1002/9781119293071.ch7
6. Q. Fang, Z. WeiJie, W. Guojun, F. Hui, Unified security architecture research for 5G wireless system, in *11th Web Information System and Application Conference*, September 2014, China
7. N. Kshetri, J. Voas, 5G, security and you. Computer **53**(3), 62–66 (2020). https://doi.org/10.1109/MC.2020.2966106
8. M. Soltani, W. Fatnassi, A. Bhuyan, Z. Rezki, P. Titus, *Physical layer Security Analysis in The Priority-Based 5G Spectrum Sharing Systems, 2019 Resilience Week (RWS)*, November 2019, USA
9. V.H. Tea, 5G subscription concealed identifier (SUCI) of IMSI based subscription protection identifier(SUPI), privacy protected with ECIES profile a protection scheme, in *IEEE Dtaport*, p. 2020
10. G. Chopra, S. Jain, R.K. Jha, Possible security attack modeling in ultra dense networks using high speed handover management, in *IEEE Transactions on Vehicular Technology*. https://doi.org/10.1109/TVT.2017.2765004
11. https://www.ccsinsight.com/blog/hyperscalers-accelerate-the-5g-edge/
12. F. Pan, H. Wen, H. Song, T. Jie, L. Wang, 5G security architecture and light weight security authentication, in *2015 IEEE/CIC International Conference on Communications in China – Workshops (CIC/ICCC)*, (2015, November). https://doi.org/10.1109/ICCChinaW.2015.7961587
13. M. Liyanage, I. Ahmad, A.B. Abro, A. Gurtov, M. Ylianttila, *Regulatory Impact on 5G Security and Privacy*, pp. 399–419. https://doi.org/10.1002/9781119293071.ch17, copy right: 2017
14. J.J.H. Wang, Wideband wide-scan millimeter-wave phased arrays for enhanced security/privacy and performance in 5G mobile wireless, in *2017 IEEE International Symposium on Antennas and Propagation & USNC/URSI National Radio Science Meeting*, July 2017. https://doi.org/10.1109/APUSNCURSINRSM.2017.8072778
15. X. Liang, X. Qiu, A software defined security architecture for SDN-based 5G network, in *2016 IEEE International Conference on Network Infrastructure and Digital Content (IC-NIDC)*, September 2016. https://doi.org/10.1109/ICNIDC.2016.7974528
16. M.A. Hasnat, S.T.A. Rumee, M.A. Razzaque, M. Mamun-Or-Rashid, Security study of 5G heterogeneous network: current solutions, limitations & future direction, in *2019 International Conference on Electrical, Computer and Communication Engineering*, February 2019, Bangladesh. https://doi.org/10.1109/ECACE.2019.8679326
17. J.-H. Noh, H.-Y. Kwon, A study on smart city security policy based on blockchain in 5G age, in *2019 International Conference on Platform Technology and Service (PlatCon)*, January 2019. https://doi.org/10.1109/PlatCon.2019.8669406
18. J.J. Cao, M. Ma, H. Li, R. Ma, Y. Sun, P. Yu, L. Xiong, A survey on security aspects for 3GPP 5G networks. IEEE Commun. Surv. Tutor **22**, 1 (2020)
19. G. Sahu, S.S. Pawar, Resource allocation using genetic algorithm in heterogeneous network, in *2019 IEEE Pune Section International Conference (PuneCon)*. https://doi.org/10.1109/PuneCon46936.2019.9105736
20. G. Chopra, R.K. Jha, S. Jain, Security issues in ultra dense network for 5G scenario, in *10th International Conference on Communication Systems & Networks (COMSNETS)*, January 2018, Bengaluru, India
21. G. Arfaoui, P. Bisson, R. Blom, R. Borgaonkar, H. Englund, E. Félix, F. Klaedtke, P.K. Nakarmi, M. Näslund, P. O'Hanlon, J. Papay, J. Suomalainen, M. Surridge, J.-P. Wary, A. Zahariev, A security architecture for 5G networks. IEEE Access **6**, 22466–22479 (2018). https://doi.org/10.1109/ACCESS.2018.2827419

22. N. Wang, P. Wang, A. Alipour-Fanid, L. Jiao, K. Zeng, Physical-layer security of 5G wireless networks for IoT: challenges and opportunities. IEEE Int. Things J. **6**(5), 8169–8181 (2019)
23. F. Liu, J. Peng, M. Zuo, Toward a secure access to 5G network, in *17th IEEE International Conference on Trust, Security and Privacy in Computing and Communications/12th IEEE International Conference on Big Data Science and Engineering (TrustCom/BigDataSE)*, New York, USA, September 2018. https://doi.org/10.1109/TrustCom/BigDataSE.2018.00156
24. A. Mazin, K. Davaslioglu, R.D. Gitlin, Secure key management for 5G physical layer security, in *IEEE 18th Wireless and Microwave Technology Conference (WAMICON)*, April 2017, USA
25. Yongpeng Wu, Ashish Khisti, Chengshan Xiao, Giuseppe Caire, Kai-Kit Wong, Xiqi Gao, A survey of physical layer security techniques for 5G wireless networks and challenges ahead IEEE J. Select. Areas Commun., Volume: 36, Issue: 4, April 2018, Page(s): 679–695
26. Li Sun, Qinghe Du, "Physical layer security with its applications in 5G networks: A review", China Commun., Volume: 14 , Issue: 12, December 2017, Page(s): 1–14, DOI: https://doi.org/10.1109/CC.2017.8246328

Software-Defined Networking-Based Ad hoc Networks Routing Protocols

G. Kirubasri, S. Sankar, Digvijay Pandey, Binay Kumar Pandey, Vinay Kumar Nassa, and Pankaj Dadheech

1 Introduction

A network provides wireless devices that communicate with each other directly without the assistance of a wireless gateway. Wireless networks often have to be managed and route data stream between wireless devices using a base station (BS) or a wireless access point (WAP) device. The design of the wireless network is divided into two respects; one is infrastructure in which the nodes are connected

G. Kirubasri · S. Sankar
Department of Computer Science and Engineering, Sona College of Technology, Salem, Tamil Nadu, India
e-mail: kirubasri.cse@sonatech.ac.in; sankar.cse@sonatech.ac.in

D. Pandey
Department of Technical Education Kanpur, IET Lucknow, Dr. A.P.J Abdul Kalam Technical University, Lucknow, India

B. K. Pandey
Department of Information Technology, College of Technology, Govind Ballabh Pant University of Agriculture and Technology, Pantnagar, Uttarakhand, India

V. K. Nassa
Department of Computer Science Engineering, South Point Group of Institutions, Sonipat, India

P. Dadheech (✉)
Computer Science & Engineering, Swami Keshvanand Institute of Technology, Management & Gramothan (SKIT), Jaipur, India

M. M. Ghonge et al. (eds.), *Software Defined Networking for Ad Hoc Networks*, EAI/Springer Innovations in Communication and Computing,
https://doi.org/10.1007/978-3-030-91149-2_5

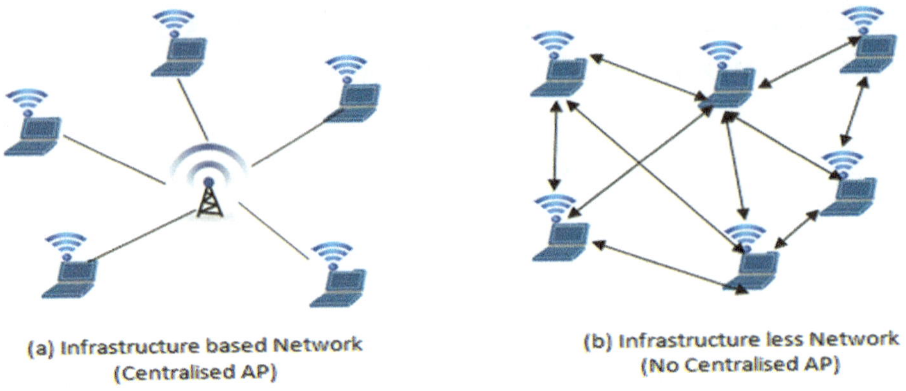

Fig. 1 Classification of wireless network

with the permanent physical network. Therefore the exchanging packets between nodes happen to employ access point (AP). The next is an infrastructure-less architecture where the nodes are connected without any fixed physical network structure [1]. Linking the endpoints to the decentralized multi-hop architecture forms the networks. The nodes can operate as routers for the sending and receipt of data due to their lack of a centralized structure. Figure 1(a) shows infrastructure-based and 1(b) shows infrastructure less network architectures of the wireless environment. The AHN is built instantaneously by connecting the devices quickly since these devices should preferably be in close contact with each other; in this setup, the connection quality and network speed will be affected when adding more devices to the network.

1.1 Design Objectives of Mobile Network (MANET) Routing Protocols

- Should be scalable.
- Should be entirely distributed, no central management.
- Should be adaptive to change in topology due to node movement.
- Calculation and maintenance of routes must include a sufficient number of nodes.
- The global exchange must be localized with a high overhead must be loop-free.
- Stagnated paths must be effectively avoided.
- Must converge fairly on best routes quickly.
- The best use of precious resources is needed: bandwidth, power of battery, memory, computation.
- The QoS should give time-sensitive traffic support assurances [1].

1.2 Properties of Mobile Networks

Mobile networks possess numerous features and the most essential are as follows:

- *Self-organizing*: Without centralized control, nodes can promote coordination among themselves on their own. With the self-organizing nature of sensor devices, there is a possibility of on-the-fly network configuration in a hostile environment.
- *Dynamic topologies*: Continually changing network setup over time, multi-hop communication takes place either in a unidirectional or bidirectional way.
- *Bandwidth constrained variable capacity links*: The dependability, efficiency, strength, and capability of wireless connections are generally lower than the wireless network.
- *Autonomous behavior*: Each node can operate as a host and router that demonstrate its independent activity. The network requires little human involvement; thus it is completely autonomous.
- *Decentralized administration*: Mobile networks differ from other networks without having prefixed infrastructure and centralized management. Mobile hosts are subject to regular network connection development and maintenance. AHN is less cost-effective [2].
- *Resource constraint devices*: In general, few or most of the mobile nodes are relying on their energy, and network devices have less memory, less power, and less lightweight attributes.
- *Shared bandwidth*: The utilization of a shared communication channel is one of the main features of wireless networks. The bandwidth for a host is low because of this sharing.
- *Vulnerable to attacks*: Because of the distributed nature of AHN and the absence of a central firewall gateway, the security risks tend to be worsened in wireless networks in routing and host configuration functionality [3].
- *Peer-to-peer communication (P2P)*: Network devices are immediately accessible through a P2P wireless link to each other's resources.
- *Distributed management*: No direct communication provision for a BS. Rather, nodes interact directly with one another, and each node is a router and a host. Failure of one node does not interfere with the overall network communication [4].

The self-sustaining nature of AHN makes them very helpful in emergencies like natural calamities, actions involving military aid, or simply to transport information rapidly between two computers. Although AHN has simplicity in its usage and scalability, the actual world has physical and performance limits. There are numerous challenges in this area without constant infrastructure. Routing, bandwidth limitations, hidden terminal issues, and small energy are the challenges of the AHN that needs to be answered. Some of the critical issues in AHN are described in the following section [5, 6].

1.3 Issues of Mobile Network

- *Dynamic network topology*: The regular node movement that causes frequent route breakdowns makes a network more challenging.
- *Insufficient admission control*: Managing the network topology and controlling the channel utilization is difficult due to the lack of centralized network authority.
- *Limited bandwidth*: The low radio frequency reduces the data rate through wireless networks. Thus, it is vital to optimizing the bandwidth by maintaining a minimal overhead [7].
- *Energy constraints*: Shorter battery life of nodes is a significant problem in the design of a network. This limited energy has to be utilized carefully for monitoring, data collection, processing, and routing packets to their destination.
- *Routing overhead*: The routing table produces stalled routes resulting in overhead routing caused by the dynamically changing devices in AHN.
- *Packet loss due to transmission error*: Wireless networks are susceptible and typically result in frequent packet loss owing to traffic crashes induced by hidden terminals, interferences, and frequent node movement interruptions.
- *Frequent network partitions*: Random node mobility leads to a partitioning of the network. This mostly impacts the middle nodes.
- *Inadequate physical security*: Mobile nodes are highly inclined to attacks both inside and outside the network.
- *Quality of service (QoS)*: Maintaining the expected QoS by the application is difficult owing to infrastructure-less network and poor link choice made by the relay nodes for routing the traffic. Reliability, resolution, throughput, packet delay, control packets, and transmission efficiency are the fundamental QoS factors required by many applications.
- *Interference*: This is the biggest concern of interference with mobile AHN. Links rely on the quality of the transmission. One node can readily interfere with another node during the transmission by which it has interfered.

Though there are some practical limitations to the overall capacity of AHN, the distributed nature gives them an advantage for a range of applications, where no middle node to limit the data forwarding and can increase network scalability over wirelessly operated networks [8, 9]. With the increased number of miniaturized hardware as well as advancement in wireless communication technology, AHN has become more popular through its widespread application. Based on the application framework the network fit into, AHN can be classified into four categories such as mobile networks (MANETs), wireless sensor networks (WSNs), wireless mesh networks (WMNs), and vehicular ad hoc networks (VANETs).

Designing an appropriate routing protocol is the most significant task in ad hoc networks and is critical for basic network operations. Nodes in AHN can communicate within the range of coverage and must depend on their neighbor to relay the packets toward the destination [10, 11]. A routing protocol should lead the packet transmission among the nodes from source to destination, and it must be wise enough to satisfy the QoS requirements and performance metrics of the associated application. Some of the exclusive properties of AHN make routine interesting and challenging. Routing is the way to find the optimal path to enable communication between nodes [12]. AHN must supply messages in the right place and an adequate method. Within AHN, each device works as a router without any common connecting access point. The user nodes must perform routing that can be mobile, unstable, and with little energy and resources [13–15].

1.4 Classification of AHN Routing

Wireless network routing techniques may be classed based on the routing information notifications method, time information use, topology information use, and resource use [16, 17]. Related to the routing information notifications method, the first category consists of proactive (table-driven) routing protocols includes destination-sequenced distance vector routing (DSDV) protocol, Wireless Routing Protocol (WRP), and cluster head gateway switch routing protocol (CGSR). The routing tables are constructed in these protocols before packets are transmitted, and each node knows the paths to all other network nodes [18, 19].

As reactive, routes are formed, and the results of the route are saved in a cache when necessary. A route repair operation is launched when intermediate nodes relocate. The protocols that come under this category are the dynamic source routing protocol (DSR), On-demand Distance Vector (AODV), associativity-based routing (ABR), flow-routing protocol (FORP), and preferred link-based routing (PLBR). The advantages of reactive and proactive routing systems are combined with hybrid routing. These protocols adapt by nature and adjust mobile nodes according to their zone and position. Core-extraction distributed routing (CEDAR), zone routing protocol (ZRP), and Zone-Based Hierarchical Link State Routing Protocol (ZHLS) come under hybrid routing [20, 21] (Fig. 2).

The protocols using the temporary information (TI) for routing are classified as past temporal and future temporal routes of information. In these routes, depending on the duration of stability of the wireless networks, each node retains the count of the beacons of its neighbors to assess temporal stability [22–25]. FORP, Route-Lifetime Assessment Based Routing (RABR), and location-based routing (LBR) are the future TI routing. DSDV, source-tree adaptive routing (STAR), DSR, AODV, fisheye state routing (FSR), and hierarchical state routing (HSR) are the past TI routing protocols. According to the use of specific resources, PAR and LAR are power-aware routing systems and location-aware routing systems, respectively [26].

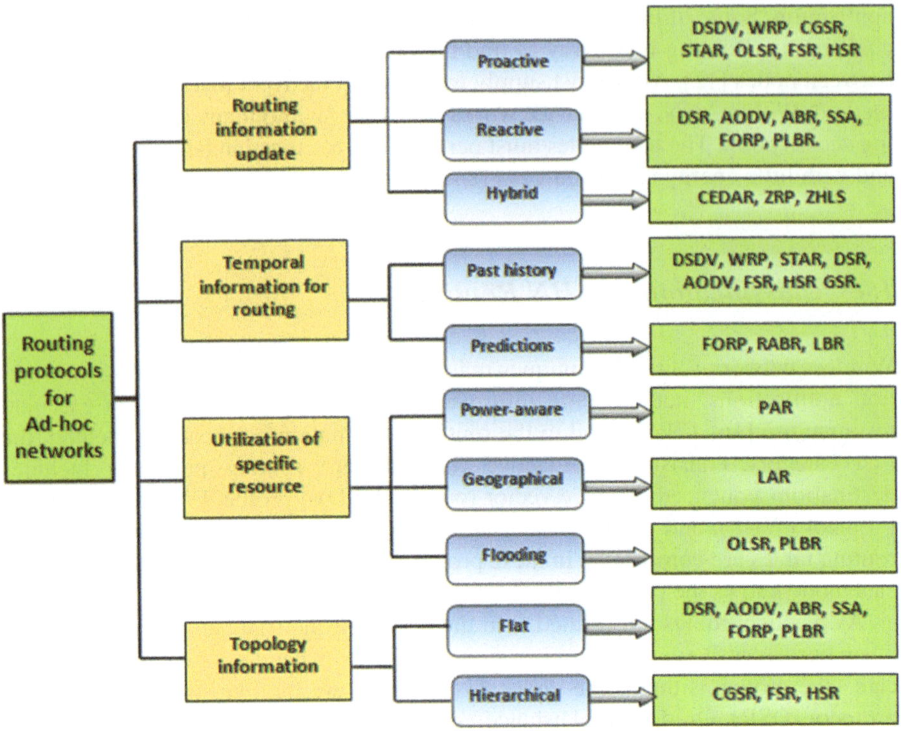

Fig. 2 Classification of routing protocols for networks

1.5 Introduction to SDN Over the Wireless Environment

The evolution of SDN has played a vital role in the creation of next-generation networks (NGN). A paradigm change has been introduced for advances in SDN, which seeks centralized designs for wired and wireless networks in a similar way [27].

SDN technique unbundles the control plane from the transmission system into switches and aggregates all control planes into a single controller. SDN allows for the management of the network behavior by software outside of the physical connection network devices. The operators can quickly create unique, distinctive new services by not connecting the hardware from the software, free from closed and proprietary platforms. The central controller collects network data from switches and calculates optimum routing pathways for switches based on global network information in an SDN-based routing architecture. The overhead routing is substantially reduced, as switches do not need to communicate routing information [28, 29].

SDN intends to create an unbounded, user-connected network administration framework for transmission devices. Depending on the scale of the network, a control plane may be one or several [30]. A rapid, dependable scattered control with distributed configuration may be established in many controller setups. Differentiated data planes from control planes play a major part in SDN in an extensive speedy computing network, whereas switches employ the flow table for data plane packet forwarding. Five important characteristics of the SDN focus are the following:

- Divide data plane from the control plane.
- Get a broad overview and provide it to the centralized controller of the whole network.
- Open interfaces between data plane devices and control plane devices.
- The network can be programmed by external applications.
- Ensure the overall management of traffic.

In a variety of infrastructure-heavy wireless network environments, SDN offers flexibility. There is a lot more of an industry turn in the infrastructure-heavy environment. This offers SDN flexibility to provide a selection of vendor appliances, enhance network latency and provide inexpensive transfers between various wireless network technologies to these dense wireless network operators. In contrast to the packet gateways (P-GWs) in the network, SDN may mainly be used for mobile networks, distributing the data plane over several inexpensive network switches.

1.6 Contributions of the Paper

The proposed paper includes the aspects of SDN-based ad hoc routing protocols as follows:

- Opportunities and challenges of SDN in AHN and impact over the design of routing protocol with SDN in AHN.
- Architectural components of SDN are explained with its functionalities related to AHN.
- Applications of SDN in different networking domains and the benefits of SDN used in various application scenarios are explained with an example.
- Classifications of SDN-based routing protocols for AHN are described with their pros and cons.

The organization of the chapter is given as follows:

In Sect. 2, comparison of SDN with traditional network, reference architecture, components of SDN, various applications of SDV over different scenario are described. Section 3 explains the classification of SDN-based networks, and Sect. 4 details the challenges of routing protocols in AHN. Section 5 presents the various SDN-based routing protocols for AHN, and Sect. 6 concludes the chapter, and the details of future enhancement are given.

2 SDN for Wireless Networks

2.1 Conventional vs Software-Defined Network

In general, the data plane and the control plane are the two components used to form the network designs. The main duty of the data plane is to send packets according to the IP address of the r recipient host. The fundamental role of the control plane is to determine the end-to-end path for forwarding the packets. In a typical network design, the operations of these planes are grouped in every router. The forwarding table is a critical component of the data plane and includes Internet Protocol (IP) address entries. The transmission function matches the arriving packet with the destination IP address in the packet header to the entries in the transmission table and determines the activities to be performed. These items in the forwarding table are configured with the routing function [31].

The networking architecture, which defines the software, removes the control plane operations from the routers and transfers them toward a piece called SDN controller (SDNC). The transmission function remains on the router, but the SDNC is used for the routing function. This distinction means that the routers are called forwarding devices. SDN is software-based, whereas conventional networking is often hardware-based. Since it is software-based, SDN is more flexible and enables users to manage resources digitally throughout the whole control plane more easily and more quickly. Conversely, conventional networks are linked and administer their setup via a switch, router, and other hardware [32].

SDNC provides application programming interface (API) connected northbound interface functionality. Due to this connection, the creators of applications may program the network directly, in contrast to utilizing the standard networking protocols [33, 34]. Instead of physical infrastructure, SDN permits customers to utilize software to prove new devices so that information technology (IT) managers may control network channels and arrange network services proactively [35]. SDN has also the capability, unlike traditional switches, to communicate better with network devices. The main distinction between SDN and conventional networking is represented by virtualization. SDN produces an abstract duplicate of the physical network when the whole network is virtualized and enables it to provide resources from a centralized place. Conversely, the physical placement of the control plane in a conventional network restricts a traffic flow control capabilities of an IT administrator.

The SDN enables the control plane to be accessible via a connected device on a software basis. This access enables IT, administrators, to regulate circulation from a centralized user interface (UI) in more detail. It provides better control over the way network functions and network configurations. This site is centralized, particularly in a network segment that has the ability to process various network settings from a centralized UI rapidly. Especially in network segmentation, the ability to process various network settings from a centralized UI rapidly. SDN has become a popular alternative to conventional networking, allowing IT managers to provide the necessary resources and bandwidth without the need for extra physical infrastructure investment. New hardware is required to boost the capacity of traditional networking.

2.2 SDN Reference Model

The SDN wireless network (SDWN), a software-based and highly efficient network control, is predicted to facilitate the administration of wireless operators. The SDWN also opens up several constraints in typical wireless networks and has new options for advancement in network structure. SDN isolates the control functions and transfers them to a central entity known as the SDN controller [36]. All nodes are used as both forwarders and end hosts in the proposed SDN-based AHN. They have restricted broadcasting ranges, such that both data transfer and control take place through multi-hop pathways. In this background the SDNC needs to perform the following tasks for efficient routing:

- For each modification in network topology, SDNC broadcasts route changes to the entire network.
- Connection details are continually sent to SDNC from which it can learn the topology of the network.
- The path to SDNC is kept in each node.

Figure 1 illustrates the three levels and many standard interfaces in SDN design. The application layer and control plane are connected through the APIs, also known as the northbound interfaces [37]. In particular, the rules and directives of the application layer for controllers are translated via the northbound interface. The boundary that connects a control unit in the control plane and hardware devices within the data plane is indicated by the southbound interface. OpenFlow, specified by the Open Networking Foundation, is a common SDN southbound interface. Several controllers, including NOX, POX, and Floodlight, have been designed based on OpenFlow. The southbound interface is used by the controller for employing the network management on the functional devices at the network layer using the southbound interface. The east-west bound interface is responsible for communication in the control layer between different controllers (Fig. 3).

In addition to other features like a load balancer, QoS, policy, and firewall, the routing feature is implemented within the SDNC as an application [38, 39]. The SDNC manages the network state, for example, network topology, link rates, and flow information. The flow-based transmission model enables QoS priority for entries (i.e., flow rules) and differential network traffic services.

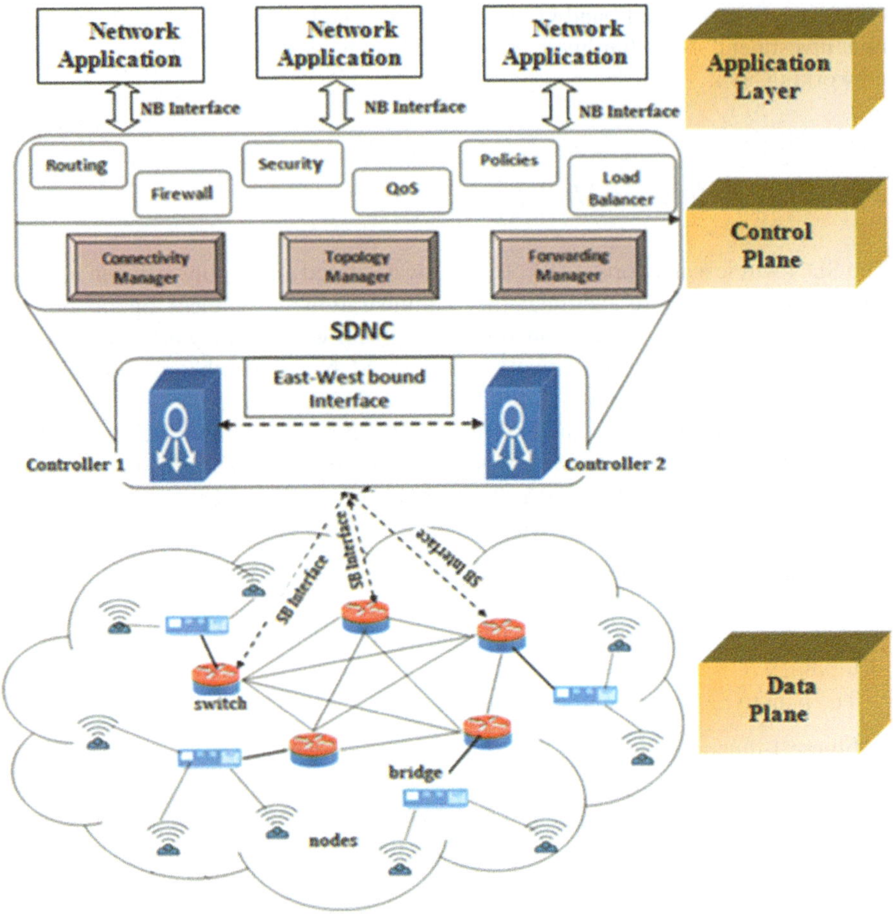

Fig. 3 SDN architecture for networks

2.3 Components of SDN Architecture

The SDN architecture consists of different components, and each has its duty in running the network effectively [40]. The basic SDN terms, SDN components are as follows:

- *Network Devices (Data Plane)*: Data plane consists of several physical and virtual connecting devices. But with SDN, just the data plane is available for network devices. So, the main *task of* these network devices is only forwarding the data. This provides a very efficient forwarding mechanism.
- *SDN Controller (Control Plane)*: The most important and central component of SDN architecture is the SDN controller, which is called the intellect of the system. The SDNC is used to govern all data plane devices. The application on the

application layer is continuously monitored. The top and the bottom layers are connected by the SDN controller using interfaces.

- *Southbound Interface*: To enable the communication with a bottom layer of network elements through southbound protocols – OpenFlow, NETCONF, OVSDB, etc.
- *Northbound Interface*: Northbound interface (NBI) is an interface of SDN and SDN controller applications, generally offering a short overview of the network's behavior and requirements straight away. This may happen at any level of abstraction and in a range of functions.
- *Network Operating System (NOS)*: It is software that is supplied on conventional server hardware separately and delivers APIs as a platform in SDN applications for switching or routing. Current examples include OpenFlow, which is being upgraded by start-ups such as ADARA, LineRate, Midokura, and Brocade, includes updates to the NOS (e.g., Big Switch, IBM, NEC, etc.). It abstracts transport layer and virtualized network services, providing northbound APIs which enable the network to be programmed and services requested by applications. In most SDN architectures, the SDN NOS stands for controlling layer.
- *Application and Services (Application Plane):* SDN applications are programs that send their requirements of network needs and network behavior to the controller explicitly, directly, and programmatically using NBIs. In addition, for internal decisions, they may use a conceptual representation of the network. An SDN system comprises one or even more NBI drivers and one SDN application logic. SDN Applications can expose an abstract network control layer by themselves to offer one or higher level NBI via the relevant NBI agents.
- The application layer is an open space to design the most inventive application by utilizing all topology details, current status, and facts about the network. Applications are implemented like network automation, network setup and administration, network monitoring, network fault resolution, and safety policies. Such SDN applications can provide final solutions for real-world enterprise networks and data centers. SDN applications are developed by network vendors.
- *Management and Administration*: In every application, there is a functional interface to a manager for the SDN controller and network element. The manager has to assign the resources to the applications residing at the higher plane from a resource pool in the lower and to provide accessibility information that allows lower and higher levels to communicate with one other. Further management features are not excluded, subject to restrictions on the exclusive control of an application, the SDN controller on any particular resource. Each unit from north to south planes may reside in a distinct administrative framework. The manager shall live in the same administrative domain as the managing entity [41].

2.4 SDN Applications

2.4.1 Data Center

Due to the versatile and vendor-neutral property of SDN network, the operators of data center, and network domain, clients are similar in using the application. Updating a policy or enhancing a specific application is not about buying a new device or confronting a proprietary physical interfacial instead programming software [42, 43]. A software-defined data center (SDDC) utilizes virtualization practices to separate the hardware infrastructure into a separate virtual machine. For SDDC customers, the benefit is that the infrastructure cannot be built; rather they simply can "rent" them via the cloud if they need computing, networking, and storage resources. Figure 4 shows the exploitation of OpenFlow for cloud and data and is referred from [44, 45]. In particular, the five primary advantages provided by SDN to the data center are as follows:

- *Processing big data*: An enterprise would like to examine big parallel information sets, but they need adequate bandwidth to do so. SDN can assist by managing performance and connectivity more efficiently.

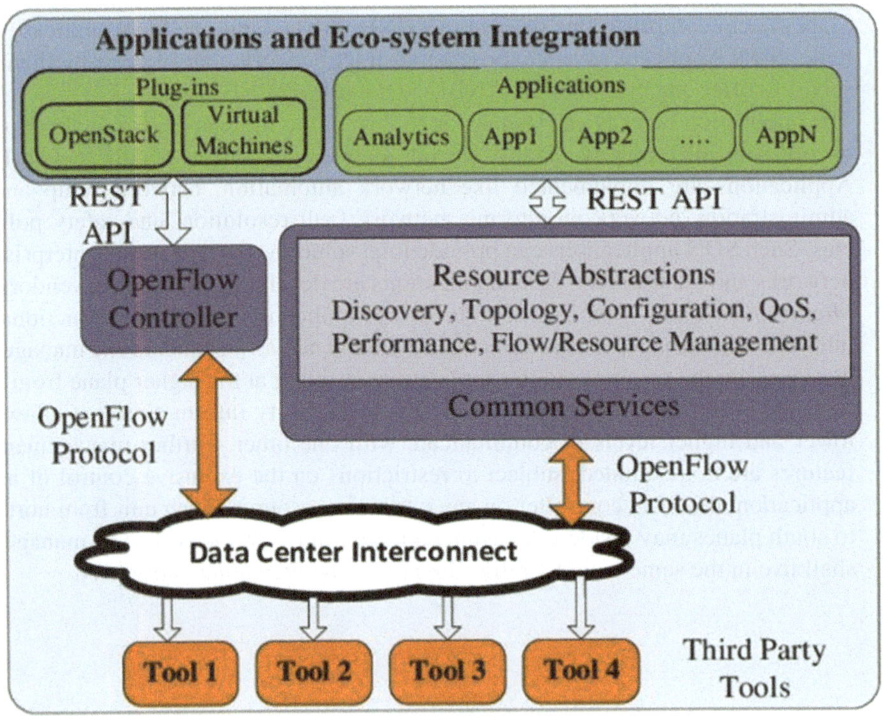

Fig. 4 The use of OpenFlow for cloud and data center

- *Cloud-based mobility support*: The emergence of the cloud is the major trend in information technology and telecommunications. Cloud is based on an idea of supplying based on demand and conscience that SDN can deliver actively depending on the availability of resources inside the data center [46].
- *Traffic management for a wide range of IP and virtual equipment*: It agrees to active routing tables, which make the priority routing easier for virtual machine based on real-time network feedback [47].
- *Flexible and adaptable network*: SDN can be used to add devices to the network more easily, reducing the risk of service disruption. SDN is more suitable to work with virtualized networks.
- *Managing policy and security*: SDN can be used to spread security policies more resourcefully and effectively across the network, including firewall devices and other key elements.

2.4.2 Telecommunications

The cellular telecommunications market is probably one of the most cost-effective. The rapidly growing numbers of mobile devices have pushed existing cellular networks to limits over the past decade. Integrating the recent developments in the current mobile architecture includes 3G Universal Mobile Telecommunications System (UMTS) and long-term evolution (LTE) with SDN that plays significant attention [48].

The greatest disadvantage of existing mobile networking topologies is the centralization of the network, handling all transit traffic by specialized equipment that includes numerous networking operations from routing to management and pricing access, resulting in increased infrastructure costs owing to device complexity and major problems with scalability. Instead, the SDN controller decides on behalf of the complete network and directs the data plane to work. The third advantage is that using SDN facilitates the launch into the telecommunications market of virtual operators and increases their competitiveness. All the suppliers are accountable for controlling their subscriber flows through their controls by virtualizing the underlying changeover equipment, without the need to pay the high for getting their infrastructure [49, 50].

2.4.3 SDN in the Industry

In comparison with conventional networking, the advantages offered by SDN led the corporate to gain attention on SDN to use it to shorten administration load and to advance services in its confidential networking systems or to create and provide marketable SDN solutions. Generally, common instances of SDN deployment in production networks is Google with its B4 network joined the realm of SDN, designed internationally to link its data centers. Google engineers indicated that the primary reason for moving to an SDN paradigm was Google's fast development of the back-end infrastructure [51].

2.4.4 Internet of Things (IoT)

IoT and SDN are the latest technologies introduced in wireless networking that create more attention in research communities due to their vibrant real-world applications. The purpose of the SDN is to connect objects via the Internet by decoupling the controller and data planes. The orchestration and control of the networks is a challenging issue for a huge, decentralized system, and there are billions of linked items. The SDN offers agility and computing to IoT networks with no compromising in conventional implementations of underlying architecture [52–54].

2.4.5 Vehicular Networks

Vehicle-to-everything (V2X) communication system in vehicular networks is a recent technology that dramatically reduces road accidents and enables high-level automation with the development of SDN [55]. Network flexibility and programmability not only change the plan of new vehicle network structures, as well the execution in the future of smart transportation systems of V2X services [56–59].

2.5 Advantages of SDN in Different Application Services

2.5.1 Security Services

The present ecosystem for virtualization supports a particular virtual service that runs at the network layer. This means that the SDN systems incorporate functions like network functions virtualization. This sort of network security enables a proactive environment to decrease risks and respond to issues extremely rapidly. Every second is critical to stop the attack whenever a violation occurs. The attack also needs to be identified, and other network components must be protected from attack. A more proactive environment can be created to respond to changes by integrating powerful services into the SDN layer.

2.5.2 Network Monitoring and Intelligence

Modern SDN technologies help to resume an important layer in the network's data center. The system architecture is very complex, and much more data than ever before must be handled. Remission, heterogeneity, and huge network traffic are the challenges that need to be reduced with a firm network supervision and intelligence layer. By integrating these technologies into the SDN architecture, it is possible to gain advantage and proper insight. Optimization, alerting, hypervisor connection, port setup, and flow may also be included in network monitoring and intelligence solutions. Such agile solutions are also employed for traffic monitoring between the cloud environment and the data center.

2.5.3 Bandwidth Management

Operators can use bandwidth management through SDN applications to ensure end-users are provided with the optimal experience of online video viewing and browsing. This SDN application may also monitor the requirements for bandwidth and provide user flows to meet Layer 7 latency and bandwidth requirements. This kind of bandwidth management approach will also improve user experience with zero buffering by improving video playback. There is little doubt at this stage that SDN in the operating networks is becoming a reality.

3 Classification of SDN-Based Networks

3.1 Mobile Networks

MANETs is a network without any assistance for infrastructure to transmit network traffic between two nodes. It is a continual self-ordered mobile device network with flat grid infrastructure and a wireless connection. It has a shared medium that is extremely demandable for radio communication. In MANET architecture, nodes or devices are often independent and act like both hosts as well as router.

MANET offers a dynamically changing topology that supports mobility [60]. The conventional MANET organization is depicted in Fig. 5.

3.2 Wireless Sensor Networks (WSNs)

A WSN uses resource-restricted sensor nodes to monitor the environment's physical circumstances, at the same time the SDN paradigm offers a straightforward and efficient communication network control method. The convergence of these systems is known as software-defined wireless sensor networks (SDWSN). Figure 6 shows a basic architecture of an SDWSN [61]. The sensor nodes in such design just transmit packets, while a conceptually centralized controller performs all control plan actions such as forwarding, service quality control, and channel assignment [62, 63]. An SDN controller can govern and optimize the performance of WSN, for example, energy consumption and communications flow, against the distributed organization of a WSN based on an integrated overview of the complete network [61].

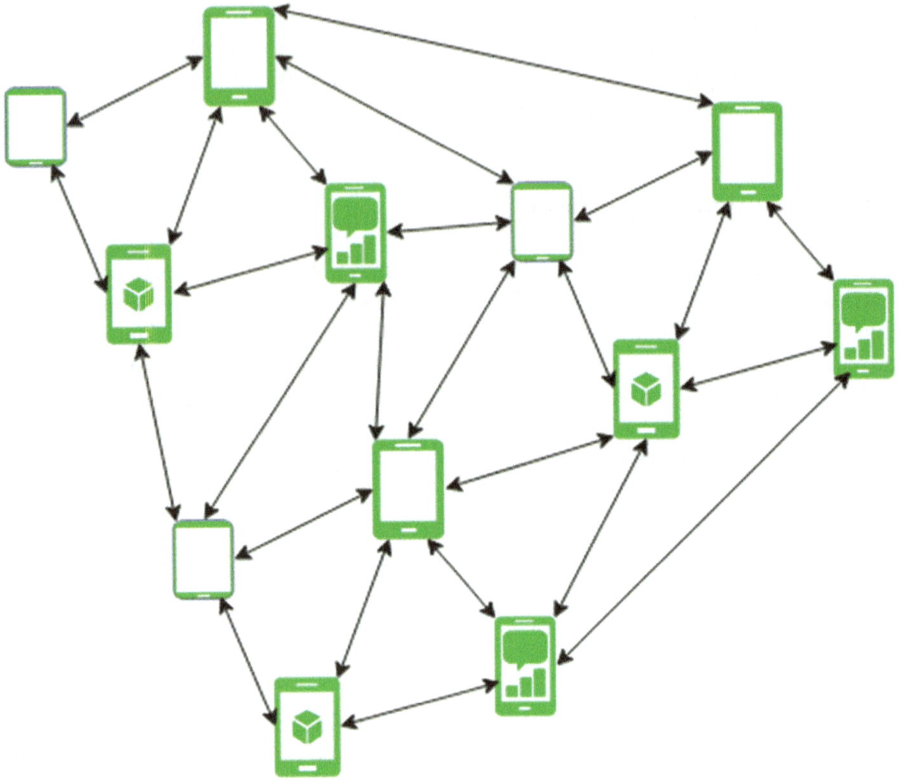

Fig. 5 MANET architecture

Fig. 6 SDN-based WSN architecture

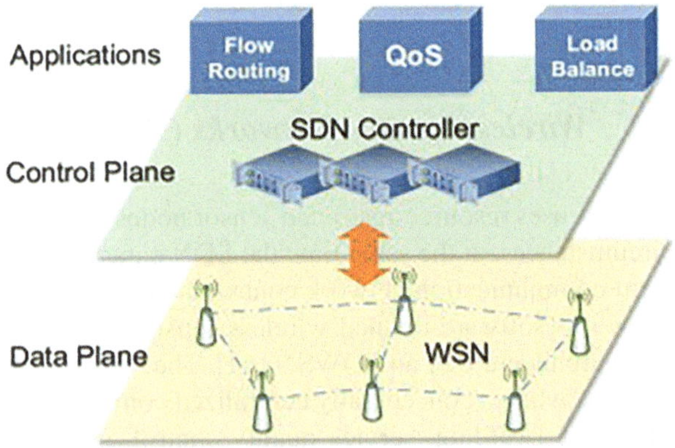

To decrease energy consumption, SDWSN employs energy management technologies like duty cycling, data aggregation in the network, and improving layer functionalities. It allows sensor nodes to be programmable to execute applications without stateless solutions like finite state machines. It features an API that makes network programming simple and versatile and allows developers to design the SDN controller in their self-knowledge and is very constructive in supervising a huge and wide range of WSN.

3.3 Wireless Mesh Networks (WMNs)

WMNs are a sort of mobile network which can organize themselves, into any network topologies dynamically. With this functionality, client devices may be connected to the network seamlessly. The design of WMNs has progressively more being implemented in recent communications and internet access applications. Naturally, it is adaptable with a variety of devices, such as switches, processors, personal computers, and routers. However, the attachment and disconnection of these nodes from the network can produce topological dynamics and fluctuation in communication needs [50].

Furthermore, limited gateway routers in WMN add to the growing congestion trouble. To mitigate these challenges, WMNs should include efficient load control, traffic engineering, and allocation of resources. In addition, multi-hop configuration causes worsening of network speed and loss of packets. Similarly, additional factors such as fluctuating connection quality, network asymmetry, and traffic loads all pose significant issues. In addition, the changing of the hardware equipment in WMN network nodes might require full or partial replacement once established and deployed, leading to substantial increases in the operational expenses. This enables WMN to be programmed to implement modifications with software applications [64].

DSR, AODV, DSDV, Optimized Link State Routing Protocol (OLSR), and Better Approach to Mobile Networking (BATMAN) are the routing protocols employed on WMNs. All these routing methods indicate a significant drop in average capacity and do not enable mobility management. The OLSR and BATMAN protocols, nonetheless, demonstrate good multi-hop performance. The performance of the OLSR protocol is often superior to the BATMAN [65, 66]. Through the application of the SDN methodology, the previous challenges of design and execution may effectively be resolved. The purpose of programmable WMNs may be achieved with a logically central controller that can remotely work and construct mesh nodes and make trouble-free data transfers. Policies for control of congestion and load balance have also been created to improve traffic and load balance management. However, the dependability of the network may be compromised by a single control, because this fault tolerance must also be taken into account during the paradigm adjustment [67].

Network operators can adopt multiple QoS policies while maintaining the requirements of the user and application [68, 69]. The proposed design in Fig. 7

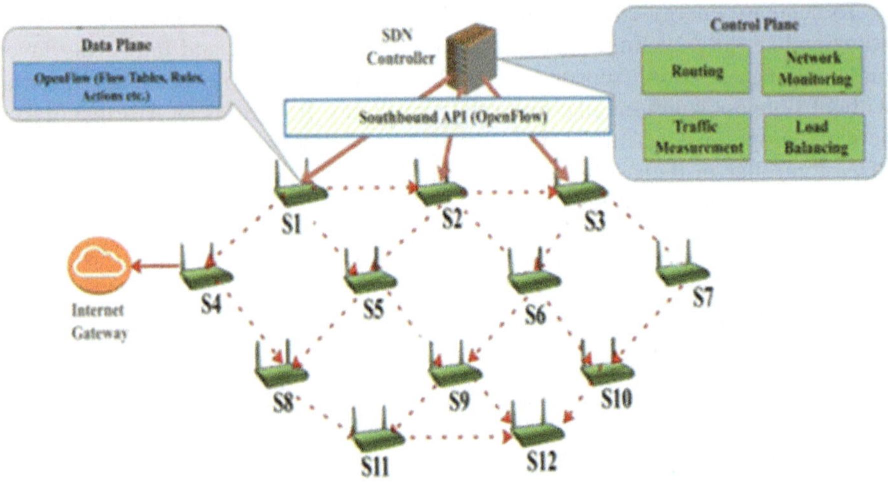

Fig. 7 SDN-based WMN architecture

demonstrates the two-stage SDN routing architecture, where the initial route from a controller to switches is identified in the first stage, and in the second the inefficient routes from the first stage are optimized. The suggested design solves additionally linkage or node failure, one of the most important wireless environment difficulties.

3.4 *Vehicular Networks (VANETs)*

The SDN-based routing framework architecture in VANETs is illustrated in Fig. 8. Three types of components are included in the entire vehicular network, i.e., the SDN controller, local controller, and forwarding nodes. Each route is also divided into several pieces of the same duration. It helps the network by optimizing the travel time between all routes between the source and destination. The SDN controller keeps up-to-date global information about the network structure. Vehicles have Wi-Fi and WiMax networking equipment, and a routing customer application for event-transmitting packets is developed [70].

The routing client uses a Wi-Fi network device to communicate data to other cars and will ask for a route from the routing server through the WiMax network device if no route entry is made for the destination. The Road Side Units (RSU), which are connected to regular clouds, function as gates on the VANET component. The burden of SDN controllers is reduced by RSUs that serve as local controllers by keeping local updated topology data in the communication area. Note that the local controller is responsible for providing mobile vehicle controls alone, but does not need the provision of automobile data. The OpenFlow protocol is designed for

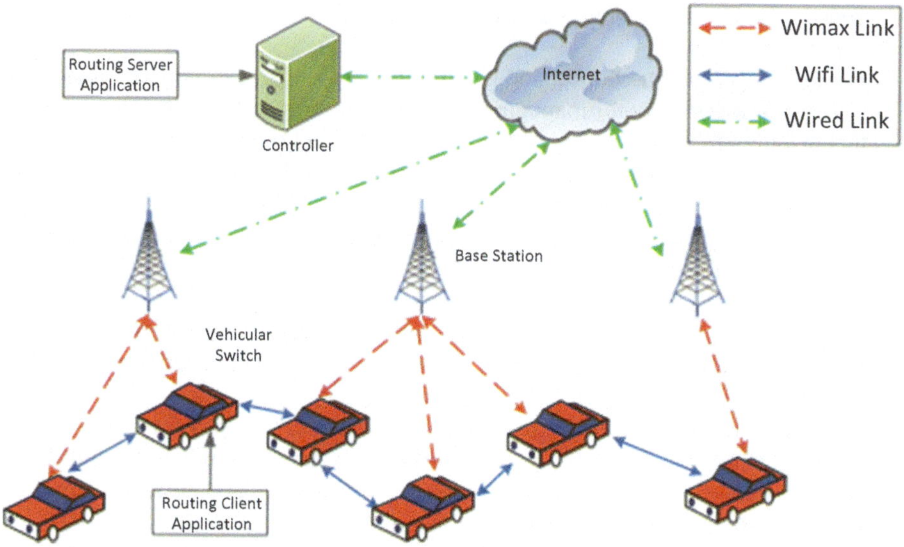

Fig. 8 SDN-based VANET architecture

enabling the interactive environment between data plane and controller, with the main advantage of efficient SDN operation of the current hardware.

4 Challenges of Network Routing Protocols

Mobility, limiting bandwidth, limiting resources, flawed medium of transfer, and location-related conflict are the major challenges for routing protocol in networks.

Current challenges in the wireless networks are given below:

- *Unavailability of the Infrastructure*: Ad hoc network that runs its nodes irrespective of any infrastructure. The mobile ad hoc network will not have permanent boundaries.
- *Lack of Centralized Monitoring Channel*: The dangers to the network are caused by the absence of a central control channel. This big ad hoc network is difficult to manage. Ad hoc network problems including transmission errors and packet loss problems are widespread.
- *Reliability and Security*: Ad hoc network has security issues and wireless connection vulnerabilities. Data transmission mistakes are also a concern with dependability owing to the constrained wireless transmission range.
- *Less Transmission Quality*: The number of fewer communication systems can lose high-rate data. The network performance might be degraded.
- *Changing the Network Topology*: In their infrastructure, nodes may roam freely. The topology of the network can be randomly limited to one end of the other

node. This might unpredictably affect the topology that degrades communications.
- *Power Consumption*: Mobile nodes in the battery-dependent network may only utilize a power supply for managing the device mobility. MANET must optimize this resource to enable nodes to connect with a broad time channel.
- *Transmission Control Protocol (TCP) Performance*: TCP is based on round trip time (RTT) measurement and network packet loss. Mobility and network congestion are impossible for TCP. Node mobility can lead to a packet loss that has a lengthy RTT.
- *Security and Privacy*: The data packet is not very secure in wireless networks. By authentication from the neighbor, a user can understand that the nearest users are hostile or friendly. Data encryption technology for packet security needs to be applied.
- *QoS Support*: In addition to the available resources, the QoS for wireless networks depends on the mobility rates of these resources. Bandwidth limits, network dynamic topology, restricted capacity processing, and storage of mobile nodes are the three basic limitations concerning service quality.
- *Energy Efficiency*: Battery-operated devices are mostly used. The microprocessor technology is trailing behind. Limiting a device's working hours indicates the necessity to save power.

5 SDN-Based Routing Protocols for AHN

5.1 Hierarchical Centralized Proactive Routing (HCPR)

A hierarchical protocol is the HCPR protocol, which builds network clusters. There is a cluster head for each cluster (CH). Nodes get information from CH about the intra-cluster routing. The HCPR protocol picks gateway nodes for inter-cluster routing besides the formation of clusters. HCPR is proactive and regularly carries out all its activities to account for network dynamics.

The HCPR functionalities, for example, the learning path to the SDNC, network architecture, and network transmission, are outlined as follows: the SDNC regularly transmits the topology discovery message (TD). A field for the radius cluster is likewise included in HCPR. For transmitting network routes, HCPR employs messages for the configuration of intra-cluster and inter-cluster routes through the route update (RU) and cluster information (CI). HCPR is a proactive protocol that updates all nodes periodically. Depending on the cluster level of a node, however, routes for inter-cluster routing can or may not exist. Nodes also invalidate their routes when link breakdowns are detected. When a node does not receive a packet from that neighbor in the last NbrMaintenance period, a link break is identified. Line breaks with the carrier sense multiple access/collision avoidance 802.11 system can also be detected by nodes if available [71].

HCPR minimizes the overhead communication by making it possible for cluster heads to set up intra-cluster routing and locate inter-cluster gateway nodes. With the use of its hierarchical routing strategy and reduced communication complexity, the results of the simulations demonstrate that the HCPR is better at achieving minimal interference than the CPR, reducing package losses.

5.2 SDN-Based Routing Protocol for Networks (SRPA)

SRPA includes the idea of SDN for path selection, and it segregates the core routing decision from the nodes and is tied to one controller. Thus, in the absence of an SRPA flow table, the node will transmit a routing request to the controller rather than relay it to neighboring nodes. This protocol uses the two data structures as SRPA flow table and SRPA neighbor table [72]. SRPA flow table for the selection of data packet paths is the main data structure. Each flow table input includes six fields: type, destination, next hop node, hop count, life span, and counter.

The neighboring table reflects the neighboring relationship between nodes. A node encapsulates and sends to the controller the neighboring table list. The controller may therefore collect topology information from the whole network. The neighboring list does not only store neighbor addresses but may also retain the weight of neighbors. To assess different parameters of a node, such as battery power and trust value, the next level of power can be employed for differing network purposes.

The experimental findings indicate that the performance of SRPA in route discovery time and RTT is superior to AODV. Many difficulties may be solved based on SRPA, such as trust and security. The weight of the neighbor list is the standard value when the node transmits the neighbor details to the controller. To increase network security, nodes can utilize the weight of the attribute to indicate their neighbor nodes' trust values. The controller may therefore evaluate confidence problems in routing decisions and acquire packet forwarding pathways that comply with security standards.

5.3 SDN- and Fog-Based VANET Routing Protocol (SFIR)

SDN-based vehicle networking system routing protocol is presented in [73]. The routing protocol is designed to find the desired route for routing data packets in real-time density. With the use of greeting messages sent by the cars, SFIR determines the traffic conditions at the fog node. The computation for a particular route is carried out using the following Eq. (1):

$$Score_i = \alpha\left(L_i + ED_i\right) + \left(\beta / D_i\right) \quad (1)$$

where $Score_i$ is the fitness value; α and β are the distance factor and the density factor, respectively; L_i is the length of the road; ED_i is the Euclidian distance; and D_i is the vehicle density of each road. The parameter value is modified according to the traffic situation value. For further investigation, the computed road conditions are subsequently forwarded to the SDN controller. The SDN controller acquires the overall position of the whole network after receiving input from all nodes. A graph based on the data obtained from the nebula takes the SDN controller to assume the road junctions as the nodes and the roads as the edges. The controller performs the Dijkstra algorithm for routing a packet, so that the shortest path may be found from source to destination.

SFR provides the uniformly dispersed traffic parts more priority. To compute road real-time distribution, the controller employs the standard deviation. The buffer size is taken into account in this protocol. A vehicle's buffer may overflow due to excessive networking. In these circumstances, they send light messages to the other cars to distribute this information. If the buffer limit drops below the threshold limit, the cars can revisit the messages. In the scenario of a genuine dispersed network, SFIR stores carry forwarding techniques. If the buffer boundary falls below the threshold, vehicles can review the messages. SFR stores carry forwarding mechanisms in the context of a scattered network.

5.4 An SDN-Based Congestion-Aware Routing Algorithm over Wireless Mesh Networks

A novel SDN control structure-based methodology to route packets is suggested [74]. Saturation to SDN and advocate for SDNR is part of the architecture of the link quality model. The network structure defined in software (SDN) promises to effectively get the network configuration and can deploy fine-grained routing algorithms with a centralized controller to fully exploit the network resources while guaranteeing the overall control over the network is acceptable. To assure network performance, the saturation link to the SDN controller is implemented, which can track the crowded path and redirect the following traffic to a non-congested road, which is the optimum in time. The benefits of SDNR are compared with normal routing techniques and shown the output.

With the OpenFlow protocol, the controller may achieve full network topology as one of SDN's important features in the design. A link quality model is necessary for the controller to identify the optimal path to the arriving fluxes from the overall network architecture. The saturation link to the SDN architecture has been set, where the controller may detect the connected state and route to the next better paths. The congested route also keeps the flows active before the packed state. This is because the identical flows will be transmitted according to the rules of transmission in the forwarders, and the controller will never be sent. Therefore the traffic on the packed track is gradually decreased, and the crowded state improves. Sometime later, the controller will again allocate more flows on that path if the path metric is detected again. The transmitters send saturation to the controller regularly. The controller takes information from all forwarders in the context of the quality template for the connection and chooses the optimum path from source to the location:

$$\left\{\begin{array}{c} Cog_{path} = 1 - \prod_{i \in path}\left(1 - Pol_i\right) \\ T_{path} = \sum_{i \in path} T_l \\ M_{path} = \left(1-\alpha\right)T_{path} + \alpha Cog_{path,} 0 \le \alpha \le 1 \end{array}\right\} \tag{2}$$

where Pol_l stands for the saturation factor of the lth link of the route, Cog_{path} indicates congested condition, and $Cog_{path} > 0$ means that the path has been larger than the congested condition. The route is seen as better if it has lesser Cog_{path} since that suggests the route has more resources available for bandwidth still. T_{path} symbolizes the wait time of the path, T_l specifies the connection latency of the path, and is a factor of a scale factor, it decides the significance of the delay and the path α. State of congestion depends on the dynamics and the network conditions. The choice of route will be more sensitive to crowded conditions for a big number of people. The selection of paths will instead be more sensitive to delay for tiny ones.

It is deployed with the OpenDaylight controller and Mininet out-of-band control framework over WMN to undertake our SDNR studies. Simulation findings indicate that SDNR performs well in average performance, packet delivery, and standardized overhead routing.

5.5 SDN-Based Geographic Routing Protocol for VANET (SDGR)

In [75] is suggested a VANET geographically supplemented SDN routing protocol named SDGR. In SDGR the geographical position and direction of every routing node are communicated regularly. The central SDN controller uses RSU to collect messages and estimate road density through an update of status messages for the vehicle. The server takes into account the width of the road in calculating the road density [76]. The server assigns weight for each route by assessing the length and density of the automobiles on the route [77]. Weight is computed based on the following equation:

$$w_{ri} = \beta * f\left(L_{ri}\right) + \frac{\gamma}{g\left(T_{ri}\right)} \tag{3}$$

In the equation, the density and length of the road are denoted by T_{ri} and L_{ri}, respectively, γ and β are the controlling parameters, and *wri* is the weight of the roadway. The shortest weight-based route computation is a lengthy operation. The

SDN controller, therefore, creates a subgraph based on the source node and the destination node from the derived graph. The process of creating subgraphs begins both from source and destination and only takes into account the related intersection. If the source node contains the destination in its routing table, the data can be sent to the destination using a greedy hop selection procedure. The vehicle takes into account speed and direction in a V2V connection the next hop selection procedure. Two primary approaches of direct mode and cross mode are applied by SDGR. The immediate mode is used to supply the data packet according to the position, speed, and direction of the next hop.

Data transmission prevents vehicles from the opposite way. As a rule, so many cars from different directions come together at the intersection that might produce various networking difficulties, such as network congestion and buffer overflow. To avoid this situation, a warning message is sent to other cars by cars if their buffer limits reach a threshold. The advantage is that packet transmission technology in junction mode can have a favorable influence on the road junction traffic signals when cars remain halted for a certain duration. The downside is that the protocol does not employ an overall strategy that minimizes the intersection of the transmission.

5.6 *Hierarchical Software-Defined Vehicular Routing (HSDVR)*

HSDVR applies in particular to SDN networks, which for any cause have broken connections with the central controller. This routing protocol provides vehicles for communication with the main controller in various clusters and forms local SDN domains [78]. With the assistance of beacon messages, the clusters are built. The beacon transmissions contain vehicle geolocation, speed, and direction. In HSDVR there are two major controllers: local and main controllers. The cluster head is regarded as a local control unit, and a major control unit is considered the principal control unit of the central architecture. The controller uses two control packets, namely, request path packet and route info packet, to transfer data from the source into the destination. A path packet is requested for the source vehicle to go to the destination route [79–87].

The route info packet is the same one in which the route request packet has been received by the destination. The host verifies if the destination route is known or not before the request path is broadcast. The host redirects packets to the target node if the host holds the details of the intended node in the routing table. If the host car has no target information, it will examine whether or not it is a controller. If the host is a controller, it will try to figure out the gates through which it may reach the target. This protocol benefits from filtering algorithms to discover the gates to the destination, which decreases the overhead transport and the drawback, which is not so good for sparse road conditions.

6 Conclusion

The emergence of SDN technology made tremendous changes in the wireless community like network design support, end-user support, programmability, routing overhead minimization, and less cost. Advancement, ease of deployment, and adverse benefits of using this in conventional network design and protocol operation lead to grabbing the attention of researchers. SDN is now playing a crucial role in all aspects and gaining more advantages in interoperability of the network, network management, security, energy efficiency, topology control, etc. The proposed article gives a clear overview of the challenges and opportunities of MANET and the types of AHN. The need for routing protocols and different categories of routing in MANET is tabulated. Classification of AHN for the application scenario was presented with architectural components. SDN-based network routing protocols for various application scenarios are explained with their pros and cons.

This work gives an insight for researchers to address the open issues in AHN. In the future, this work can be expanded with more routing parameters and application need for categorizing the routing protocols proposed for SDN-based routing in AHN. Application-specific requirements for SDN for routing protocols design in WSN will be described to give a clear overview of researchers to identify the problems in design aspects.

References

1. D. Helen, D. Arivazhagan, Applications, advantages, and challenges of ad-hoc networks. J. Acad. Ind Res. (JAIR) **2**(8), 453–457 (2014)
2. N. Raza, M.U. Aftab, M.Q. Akbar, O. Ashraf, M. Irfan, Mobile networks applications and their challenges. Commun. Netw. **8**(3), 131–136 (2016)
3. M.A. Pandey, Introduction to mobile ad hoc network. Int. J. Sci. Res. Publ. **5**(5), 1–6 (2015)
4. M. Chitkara, M.W. Ahmad, Review on manet: Characteristics, challenges, imperatives, and routing protocols. Int. J. Comput. Sci. Mob. Comput. **3**(2), 432–437 (2014)
5. G. Kirubasri, U. Maheswari, R. Venkatesh, A survey on hierarchical cluster-based routing protocols for wireless multimedia sensor networks. J. Converg. Inf. Technol. **9**(6), 19 (2014)
6. S. Sennan, R. Somula, A.K. Luhach, G.G. Deverajan, W. Alnumay, N.Z. Jhanjhi, et al., Energy efficient optimal parent selection based routing protocol for Internet of Things using firefly optimization algorithm. Transac. Emerg. Telecommun. Technol. **32**, e4171 (2020)
7. H. Zemrane, Y. Baddi, A. Hasbi, Mobile ad hoc networks for intelligent transportation system: Comparative analysis of the routing protocols. Proc. Comp. Sci. **160**, 758–765 (2019)
8. V.K. Mishra, A. Dusia, A. Sethi, *Routing in Software-Defined Mobile Ad Hoc Networks (Sd-Manet)*. (US Army Research Laboratory Aberdeen Proving Ground United States, 2018)
9. S. Sennan, S. Ramasubbareddy, S. Balasubramaniyam, A. Nayyar, M. Abouhawwash, N.A. Hikal, T2FL-PSO: Type-2 fuzzy logic-based particle swarm optimization algorithm used to maximize the lifetime of internet of things. IEEE Access **9**, 63966–63979 (2021)
10. G. Kirubasri, N.U. Maheswari, A study on hardware and software link quality metrics for wireless multimedia sensor networks. Int. J. Adva. Net. Appl. **8**(3), 3103 (2016)

11. S. Sennan, S. Balasubramaniyam, A.K. Luhach, S. Ramasubbareddy, N. Chilamkurti, Y. Nam, Energy and delay aware data aggregation in routing protocol for internet of things. Sensors **19**(24), 5486 (2019)
12. A.R. Rajeswari, in *Recent Trends in Communication Networks*, A Mobile Ad Hoc Network Routing Protocols: A Comparative Study (IntechOpen, 2020)
13. G. Kirubasri, A contemporary survey on clustering techniques for wireless sensor networks. Turk. J. Comp. Mathe. Educ. (TURCOMAT) **12**(11), 5917–5927 (2021)
14. S. Sankar, P. Srinivasan, A.K. Luhach, R. Somula, N. Chilamkurti, Energy-aware grid-based data aggregation scheme in routing protocol for agricultural internet of things. Sustain. Comput. Inform. Syst. **28**, 100422 (2020)
15. G. Kirubasri, N.U. Maheswari, R. Venkatesh, Novel energy efficient predictive link quality based reliable routing for wireless multimedia bio-sensor networks in bio-medical invention research and bionic utilities monitoring application. Int. J. Biomed. Eng. Technol. **26**(3–4), 219–236 (2018)
16. S. Rani, S.H. Ahmed, Multi-Hop Routing in Wireless Sensor Networks: An Overview, Taxonomy, and Research Challenges (2015)
17. A.S. Navaz, D.G.K. Nawaz, Layer orient time domain density estimation technique based channel assignment in tree structure wireless sensor networks for fast data collection. Int. J. Eng. Technol. **8**(3), 1506–1512 (2016)
18. S. Samanta, S.S. Singhar, A.H. Gandomi, S. Ramasubbareddy, S. Sankar, A WiVi based IoT framework for detection of human trafficking victims kept in hideouts, n International Conference on Internet of Things (pp. 96–107). (Springer, Cham, 2020)
19. T.K. Saini, S.C. Sharma, Prominent unicast routing protocols for mobile ad hoc networks: criterion, classification, and key attributes. Ad Hoc Netw. **89**, 58–77 (2019)
20. G. Kirubasri, A machine learning model for improved prediction of Alzheimer's progression. Int. J. Adv. Sci. Technol. **29**(6), 4204–4215 (2020)
21. A.R. Ragab, A new classification for network. iJIM **14**(14), 215 (2020)
22. A. Boukerche, B. Turgut, N. Aydin, M.Z. Ahmad, L. Bölöni, D. Turgut, Routing protocols in ad hoc networks: a survey. Comput. Netw. **55**(13), 3032–3080 (2011)
23. S. Sankar, P. Srinivasan, Fuzzy sets based cluster routing protocol for internet of things. Int. J. Fuz. Syst. Appl. IJFSA **8**(3), 70–93 (2019)
24. S. Sankar, P. Srinivasan, Enhancing the mobility support in internet of things. Int. J. Fuz. Syst. Appl. IJFSA **9**(4), 1–20 (2020)
25. M.G. Kirubasri, N. UmaMaheswari, R. Venkatesh, A robust intra-cluster communication for wireless multimedia sensor networks using link quality analysis. Int. J. Pure Appl. Math. **117**(8), 149–154 (2017)
26. P. Misra, Routing protocols for ad hoc mobile wireless networks. *Courses Notes* (1999). Available at http://www.cis.ohio-state.edu/~jain/cis788-99/adhoc_routing/index.html
27. W. Xia, Y. Wen, C.H. Foh, D. Niyato, H. Xie, A survey on software-defined networking. IEEE Commun. Surv. Tutor. **17**(1), 27–51 (2014)
28. B. Mishra, D. Jena, R. Somula, S. Sankar, Secure key storage and access delegation through cloud storage. Int. J. Knowl. Syst. Sci. (IJKSS) **11**(4), 45–64 (2020)
29. Z.J. Han, W. Ren, A novel wireless sensor networks structure based on the SDN. Int. J. Distribut. Sensor Netw. **10**(3), 874047 (2014)
30. L.F. da Silva Santos, F.F. de Mendonça Júnior, K.L. Dias, μSDN: an SDN-based routing architecture for wireless sensor networks, in *2017 VII Brazilian Symposium on Computing Systems Engineering (SBESC)*, (IEEE, 2017, November), pp. 63–70
31. M. Hadley, D. Nicol, R. Smith, Software-defined networking redefines performance for ethernet control systems, in *Power and Energy Automation Conference*, (2017, March)
32. D. Kreutz, F.M. Ramos, P.E. Verissimo, C.E. Rothenberg, S. Azodolmolky, S. Uhlig, Software-defined networking: a comprehensive survey. Proc. IEEE **103**(1), 14–76 (2014)
33. S. Sankar, P. Srinivasan, Multi-layer cluster based energy aware routing protocol for internet of things. Cyber. Inform. Technol. **18**(3), 75–92 (2018)

34. G. Kirubasri, Energy efficient routing using machine learning based link quality estimation for WMSNs. Turk. J. Comput. Math. Educ. (TURCOMAT) **12**(11), 3767–3775 (2021)
35. M. Jammal, T. Singh, A. Shami, R. Asal, Y. Li, Software defined networking: state of the art and research challenges. Comput. Netw. **72**, 74–98 (2014)
36. C.Y. Hans, G. Quer, R.R. Rao, Wireless SDN mobile ad hoc network: from theory to practice, in *2017 IEEE International Conference on Communications (ICC)*, (IEEE, 2017, May), pp. 1–7
37. K. Poularakis, Q. Qin, K.M. Marcus, K.S. Chan, K.K. Leung, L. Tassiulas, Hybrid sdn control in mobile ad hoc networks, in *2019 IEEE International Conference on Smart Computing (SMARTCOMP)*, (IEEE, 2019, June), pp. 110–114
38. S. Sankar, P. Srinivasan, Internet of things (iot): a survey on empowering technologies, research opportunities and applications. Int. J. Pharm. Technol. **8**(4), 26117–26141 (2016)
39. K. Poularakis, G. Iosifidis, L. Tassiulas, SDN-enabled tactical ad hoc networks: extending programmable control to the edge. IEEE Commun. Mag. **56**(7), 132–138 (2018)
40. V.S. Shukla, SDN transport architecture and challenges, in *Optical Fiber Communication Conference*, (Optical Society of America, 2015, March), pp. W4J-1
41. S. Schaller, D. Hood, Software defined networking architecture standardization. Comput. Stand. Interf. **54**, 197–202 (2017)
42. S. Sankar, P. Srinivasan, S. Ramasubbareddy, B. Balamurugan, Energy-aware multipath routing protocol for internet of things using network coding techniques. Int. J. Grid Utility Comput. **11**(6), 838–846 (2020)
43. https://www.opennetworking.org/sdn.resources/sdn-definition
44. S. Kiruthika, G. Kirubasri, Sentiment analysis for product improvement in e-commerce sites – a simulation. Int. J. Adv. Sci. Technol. **29**(06), 4245–4252 (2020)
45. https://www.datacenterknowledge.com/archives/2013/07/26/7-software-defined-networking-considerations/
46. S. Sankar, P. Srinivasan, Energy and load aware routing protocol for internet of things. Int. J. Adv. Appl. Sci. (IJAAS) **7**(3), 255–264 (2018)
47. M. Liyanage, A. Gurtov, Ylianttila, M. (Eds.), *Software Defined Mobile Networks (SDMN): Beyond LTE Network Architecture* (Wiley, 2015)
48. E.M. Royer, C.-K. Toh, A review of current routing protocols for ad hoc mobile wireless networks by EM Royer, CK Toh in IEEE Personal communications, 1999. IEEE Pers. Commun. **6**(2), 46–55 (1999)
49. S. Kiruthika, G. Kirubasri, Improving the efficiency of a dual corpus text to speech synthesis system using a Prefetch buffer. Int. J. Adv. Sci. Technol. **29**(06), 4253–4258 (2020)
50. T. Bakhshi, State of the art and recent research advances in software defined networking, in *Wireless Communications and Mobile Computing*, (2017)
51. S.K. Tayyaba, M.A. Shah, O.A. Khan, A.W. Ahmed, Software defined network (sdn) based internet of things (iot) a road ahead, in *Proceedings of the International Conference on Future Networks and Distributed Systems*, (2017, July), pp. 1–8
52. L. Nkenyereye, L. Nkenyereye, S.M. Islam, Y.H. Choi, M. Bilal, J.W. Jang, Software-defined network-based vehicular networks: a position paper on their modeling and implementation. Sensors **19**(17), 3788 (2019)
53. S. Sankar, R. Somula, R.L. Kumar, P. Srinivasan, M.A. Jayanthi, Trust-aware routing framework for internet of things. Int. J. Knowl. Syst. Sci. (IJKSS) **12**(1), 48–59 (2021)
54. S. Sankar, S. Ramasubbareddy, F. Chen, A.H. Gandomi, Energy-efficient cluster-based routing protocol in internet of things using swarm intelligence, in *2020 IEEE Symposium Series on Computational Intelligence (SSCI)*, (IEEE, 2020, December), pp. 219–224
55. S. Sennan, S. Ramasubbareddy, A.K. Luhach, A. Nayyar, B. Qureshi, CT-RPL: cluster tree based routing protocol to maximize the lifetime of internet of things. Sensors **20**(20), 5858 (2020)
56. S. Sankar, P. Srinivasan, Internet of things based digital lock system. J. Comput. Theor. Nanosci. **15**(9–10), 2758–2763 (2018)

57. S. Sankar, P. Srinivasan, R. Saravanakumar, Internet of things based ambient assisted living for elderly people health monitoring. Res. J. Pharm. Technol. **11**(9), 3900–3904 (2018)
58. S. Sankar, P. Srinivasan, Composite metric based energy efficient routing protocol for internet of things. Int. J. Intell. Eng. Syst. **10**(5), 278–286 (2017)
59. A. Hakiri, A. Gokhale, P. Berthou, D.C. Schmidt, T. Gayraud, Software-defined networking: challenges and research opportunities for future internet. Comput. Netw. **75**, 453–471 (2014)
60. S. Mirza, S.Z. Bakshi, Introduction to MANET. Int. Res. J. Eng. Technol. **5**(1), 17–20 (2018)
61. Q. Liu, L. Cheng, R. Alves, T. Ozcelebi, F. Kuipers, G. Xu, et al., Cluster-based flow control in hybrid software-defined wireless sensor networks. Comput. Netw. **187**, 107788 (2021)
62. M. Rezaee, M.H.Y. Moghaddam, SDN-based quality of service networking for wide area measurement system. IEEE Transac. Indust. Inform. **16**(5), 3018–3028 (2019)
63. A.A. Abdelltif, E. Ahmed, A.T. Fong, A. Gani, M. Imran, SDN-based load balancing service for cloud servers. IEEE Commun. Mag. **56**(8), 106–111 (2018)
64. S.Y. Shahdad, A. Sabahath, R. Parveez, Architecture, issues and challenges of wireless mesh network, in *2016 International Conference on Communication and Signal Processing (ICCSP)*, (IEEE, 2016, April), pp. 0557–0560
65. T. Clausen, P. Jacquet, C. Adjih, A. Laouiti, P. Minet, P. Muhlethaler, ... L. Viennot, *Optimized link state routing protocol (OLSR)*, (2003)
66. D. Johnson, N.S. Ntlatlapa, C. Aichele, *Simple Pragmatic Approach to Mesh Routing Using BATMAN*, (2008)
67. I.T. Haque, N. Abu-Ghazaleh, Wireless software defined networking: A survey and taxonomy. IEEE Commun. Surv. Tutor. **18**(4), 2713–2737 (2016)
68. V. Nascimento, M. Moraes, R. Gomes, B. Pinheiro, A. Abelém, V.C. Borges, et al., Filling the gap between software defined networking and wireless mesh networks, in *10th International Conference on Network and Service Management (CNSM) and Workshop*, (IEEE, 2014, November), pp. 451–454
69. D.B. Rawat, S. Reddy, Recent advances on software defined wireless networking, in *SoutheastCon 2016*, (IEEE, 2016, March), pp. 1–8
70. H. Trivedi, S. Tanwar, P. Thakkar, Software defined network-based vehicular ad hoc networks for intelligent transportation system: recent advances and future challenges, in *International Conference on Futuristic Trends in Network and Communication Technologies*, (Springer, Singapore, 2018, February), pp. 325–337
71. A. Dusia, *Software-Defined Architecture and Routing Solutions for Mobile Ad hoc Networks* (Doctoral dissertation, University of Delaware, 2019)
72. D. Wei, Z. Liu, H. Cao, SRPA: SDN-based routing protocol for ad hoc networks, in *2018 9th International Conference on Information Technology in Medicine and Education (ITME)*, (IEEE, 2018, October), pp. 1012–1017
73. N. Noorani, S.A.H. Seno, Routing in VANETs based on intersection using SDN and fog computing, in *2018 8th International Conference on Computer and Knowledge Engineering (ICCKE)*, (IEEE, 2018, October), pp. 339–344
74. H. Fu, Y.A. Liu, K.M. Liu, Y.Y. Fan, An SDN-based congestion-aware routing algorithm over wireless mesh networks, in *Wireless Communication and Sensor Network: Proceedings of the International Conference on Wireless Communication and Sensor Network (WCSN 2015)*, (2016), pp. 111–119
75. X. Ji, H. Yu, G. Fan, W. Fu, SDGR: an SDN-based geographic routing protocol for VANET, in *2016 IEEE International Conference on Internet Of Things (iThings) and IEEE Green Computing and Communications (GreenCom) and IEEE Cyber, Physical and Social Computing (CPSCom) and IEEE Smart Data (SmartData)*, (IEEE, 2016, December), pp. 276–281
76. R.A. Nazib, S. Moh, A comparative study on routing protocols for vehicular ad hoc networks based on software defined networking
77. S. Sankar, P. Srinivasan, Mobility and energy aware routing protocol for healthcare IoT application. Res. J. Pharm. Technol. **11**(7), 3139–3144 (2018)

78. S. Correia, A. Boukerche, R.I. Meneguette, An architecture for hierarchical software-defined vehicular networks. IEEE Commun. Mag. **55**(7), 80–86 (2017)
79. G. Ravi, K.R. Kashwan, A new routing protocol for energy efficient mobile applications for ad hoc networks. Comput. Elect. Eng. **48**, 77–85. https://doi.org/10.1016/j.compeleceng.2015.03.023,2015. (ECE), Cited - 29
80. M. Usha, P. Kavitha, Anomaly based intrusion detection for 802.11 networks with optimal features using SVM classifier. Wireless Netw. **23**(8), 2431–2466 (2017). https://doi.org/10.1007/s11276-016-1300-5. Cited - 16
81. S. Sennan, S. Ramasubbareddy, S. Balasubramaniyam, A. Nayyar, C.A. Kerrache, M. Bilal, MADCR: mobility aware dynamic clustering-based routing protocol in internet of vehicles. China Commun. **18**(7), 69–85
82. D.S. Nayagi, G.G. Sivasankari, V. Ravi, K.R. Venugobal, S. Sennan, REERS: reliable and energy efficient route selection algorithm for heterogeneous Internet of things applications. Int. J. Commun. Syst. **34**(13), e4900 (2021)
83. M.S. Kumar, S. Sankar, V.K. Nassa, D. Pandey, B.K. Pandey, W. Enbeyle, Innovation and creativity for data mining using computational statistics, in *Methodologies and Applications of Computational Statistics for Machine Intelligence*, (IGI Global, 2021), pp. 223–240
84. T.K. Revathi, B. Sathiyabhama, S. Sankar, A deep learning based approach for diagnosing coronary inflammation with multi-scale coronary response dynamic balloon tracking (MSCAR-DBT) based artery segmentation in coronary computed tomography angiography (CCTA). Ann. Rom. Soc. Cell Biol. **25**(6), 4936–4948 (2021)
85. T.K. Revathi, B. Sathiyabhama, S. Sankar, Diagnosing cardio vascular disease (CVD) using generative adversarial network (GAN) in retinal fundus images. Ann. Rom. Soc. Cell Biol., 2563–2572 (2021)
86. S. Sennan, S. Ramasubbareddy, A. Nayyar, Y. Nam, M. Abouhawwash, LOA-RPL: novel energy-efficient routing protocol for the internet of things using lion optimization algorithm to maximize network lifetime. Comput. Mat. Contin. **61**(1) (2021)
87. M.S. Karthiprem, S. Selvarajan, M.S. Sankar, Recognizing the moving vehicle while driving on Indian roads. Int. J. Appl. Eng. Res. **10**(20), 41471–41477 (2015)

Fuzzy Approach-Based Stable Energy-Efficient AODV Routing Protocol in Mobile Ad hoc Networks

Shubham Choudhary, Vipul Narayan, Mohammad Faiz, and Sabyasachi Pramanik

1 Introduction

In recent years wireless and networking field has experienced considerable improvements. Ad hoc is derived from a Latin word that means "for this purpose." Word MANET [1] is an assembly of independent nodes which communicate via radio waves through one or more intermediate nodes. MANET is a distributed network that does not have any infrastructure. Fast deployment is the main feature of mobile Ad hoc networks; this key makes MANET proper for various types of critical environments. Mobile ad hoc network has lots of applications because of its dynamic and flexible nature such as emergency operation, disaster relief, military operations, vehicular network, causal meeting, etc. MANETs are gaining fame, and their real-time and multimedia applications are rising. Mobile ad hoc network has a lot of challenges such as dynamic topology, multi-hop routing, hidden terminal problem, exposed terminal problem, packet loss, mobility, and security threat. Routing protocol defines the set of instructions that directs data packets from one node to another node. A flooding mechanism is used for broadcasting the RREQ packet to determine the possible routes from the source to destination nodes without bothering network statuses such as energy, bandwidth, and link quality.

In MANET stability-based routing, protocol selects the enduring path. There are different types of parameters that are used for the estimation of link stability like signal strength, link expiration time, and relative speed between nodes. Route stability issue emphasis on following features.

S. Choudhary (✉) · V. Narayan (✉) · M. Faiz
Department of CSE, M. M. M University of Technology, Gorakhpur, India

S. Pramanik
Department of CSE, Haldia Institute of Technology, Haldia, India

M. M. Ghonge et al. (eds.), *Software Defined Networking for Ad Hoc Networks*, EAI/Springer Innovations in Communication and Computing,
https://doi.org/10.1007/978-3-030-91149-2_6

1.1 Stable Routes

To reduce the traffic latency and for maximizing the network throughput, a reliable source node to destination node connection must be guaranteed. A route selection is done based on node motion and probability modal of future path availability.

1.2 Efficient Route Repair

If there is a service interruption due to failure of route, this interruption can be avoided by the creation of a substitute path before the breakage of the current path.

1.3 Connectivity of Network

Connectivity and topology attributes are decided by the link dynamics, and their fundamental issues to network design stability of link can be defined as how long it can sustain communication between source nodes to the destination node. Fuzzy logic is an approach which not only supports several inputs, but it is also an appropriate choice for solving multi-metrics problem in the network.

1.4 Route Maintenance

In this phase when there is a detection of link breaks for the next node or the next node becomes unreachable, then the destination node sends a RERR message to the source node.

2 Related Work

In [2], the author proposed "fuzzy logic-based wireless multipath routing" in which only the hop count metric is used for path selection and message forwarding. Network status is taken as a key feature for path selection which differs from excellent to poor and receipts excellent factor for the path selection.

In [3], each node has a mobile node table based on bandwidth and delay in reaching the neighbor. This type of table is recognized as a neighbor table. Three extra parameters are added in RREQ packet named as the min bandwidth, min LET (link expiration time), and delay sum. Based on these parameters, fuzzy cost is evaluated, and the optimal route is selected based on the fuzzy cost.

In [4], "fuzzy stochastic multipath routing (FSMR)" chooses multiple metrics like hop count (HC), residual energy (RE), and signal strength. The selection of the optimal path is based on the parameters like energy consumption rate at the node, count of the intermediate node, and buffer occupancy rate. Based on these parameters, optimal path is selected, and the fuzzy cost is calculated.

In [5], the "multi-objective pareto-optimal" protocol uses a genetic algorithm (GA). In this technique, five quality of services (QoS) parameters are taken for stable path selection. The QoS parameters are delay, bandwidth, packet loss rate, jitter, and blocking property. Based on the following parameter, fuzzy cost is calculated, and by using fuzzy cost, stable route is selected.

In [6], the author proposed genetic algorithm is superior to the normal routing algorithm. It considers energy consumption rate, link stability, and hop count. This tactic works better in the mobility model. An entropy-based model has been given for the stability of routes in MANETs [7]. The combined value of entropy, QoS, and fuzzy logic is used for better path selection.

A fuzzy scheduler algorithm [8] priority index is calculated for each data packet. It also considers the priority index while selecting the input. It takes three input variables and one output variable. Inputs are given for fuzzification, and then with the help of *IF-THEN* rules, output is aggregated, and the defuzzification method is applied to get the crisp value.

The author proposed a scheme in which there are two fuzzy controllers [9]. The fuzzy controller takes three input parameters packet queue occupancy, number of intermediate nodes, and distance between internodes. The second controller has only one parameter and one output; this calculates network lifetime from the source node to the destination node. Different performance metrics are improved by this scheme.

Bandwidth delay constraint and Dijkstra's algorithm are used in the fuzzy logic-based algorithm for finding the shortest path [10]. Two parameters bandwidth and delay are used for QoS constraints. Improved rank-based multipath routing (IRMP) [11] considers five parameters for QoS constraint like power consumption, computing efficiency, bandwidth, number of the intermediate node, and traffic load. Weighted round robin (WRR) architecture is used for packet forwarding [12]. Scheduler queues are served according to weight. Bandwidth and delay metrics are also considered for the optimal route selection. Artificial fish swarm (AFS) algorithm considers bandwidth, delay, and jitter as quality of services (QoS) [13] and applies the fuzzy logic for the simulation result.

3 AODV Routing Protocol

Ad hoc on-demand distance vector routing (AODV) [14] is a responsive protocol. It is a self-configurable, dynamic, and multi-hop routing algorithm. It supports unicast and multicast routing. It uses hop count to select the path. It uses the flooding concept for discovering the route. It transmits the route request (RREQ) packet to the neighbor node which is in the transmission range. Neighbor nodes check whether a node is fresh or itself destination then reply in a unicast manner. For the selection of route, minimum hop count metrics are used. Any intermediate node rebroadcasts the route request (RREQ) packet if it is not the destination node after the increment of the hop count value by one. Multiple route request (RREQ) packets are received by an intermediate node with a similar identification number having a different hop count. Node checks the route request packet separately; if the value of hop count is less than the previously received RREQ packets, then it updates the reverse table otherwise discard the route request (RREQ) packet. The intermediate nodes may broadcast identical identification RREQ packets more than once. In the AODV protocol, energy is consumed, network bandwidth is wasted, and network traffic is increased.

4 Fuzzy Logic System

The fuzzy logic theory was suggested by Zadeh in 1995 [15]. Fuzzy Logic can be defined as a "degree of truthness." The idea of partial truth, where the truth value ranges between completely true and completely false, came from fuzzy logic. FLS gives the relation between the crisp input and output variable using the IF-THEN rule, and IF-THEN rules are designed by fuzzy system designers [16].

4.1 *Fuzzification*

Fuzzification takes the input and maps the crisp value into proper fuzzy logic values using the membership function. The membership function is related to each element. μA represents the membership function that lies between [0, 1].

Where μA(Xi): membership degree of Xi in the fuzzy set.

Thresholds are used for the operation of the system. The activation point is denoted by the lower threshold, and the model operates within the upper and lower threshold region [14].

Crisp input is made fuzzy by a fuzzifier. There are multiple types of membership functions. Equation of triangular and trapezoidal membership function is given below:

$$\text{Triangle}\left(\left(x,a,b,c\right)\right)=\begin{cases} 0 & x \le a_1, \\ \dfrac{x-a_1}{b_1-a_1} & a_1 \le x \le b_1 \\ \dfrac{c_1-x}{c_1-b_1} & b_1 \le x \le c_1 \\ 0 & x \ge c_1 \end{cases} \tag{1}$$

$$\text{Trapezoid}\left(\left(x,a,b,c\right)\right)=\begin{cases}0 & x\le a_2,\\ \dfrac{x-a_2}{b_2-a_2} & a2\le x\le b2\\ 1 & c2\le x\le d2\\ 0 & x\ge d_2\end{cases} \tag{2}$$

4.2 *Fuzzy Inference Engine*

A rule base is used in a fuzzy logic system which is a set of rules that defines the behavior of the fuzzy decision engine. Fuzzy rules depend on both the number of input and the membership function of each input. In fuzzy rules X part is antecedents, and variable Y is the consequent part of fuzzy logic. A fuzzy rule-based system contains at most ($l \times m \times k$) rules.

Where, l, m, and k: number of input variables.

4.3 *Defuzzification*

Defuzzification [14] in fuzzy logic system is a conversion of fuzzy set into a single crisp value. It uses the weighted mean approach to extract the crisp value. Centroid method (CM), center of sum method (COS), and mean of maxima (MOM) method are the popular defuzzification methods. The mathematical expression for centroid is given below [17]:

$$COG=\frac{\int \mu_A\left(x\right).xdx}{\int \mu_A\left(x\right).dx}$$

where μ_A: weight of the associated output membership function.

5 Proposed Approach

In traditional AODV single metric is used for the optimal path selection. This may not be enough metrics for the path selection. For better route selection, fuzzy logic considers different parameters hop count, residual energy, speed, link expiration time, and bandwidth. It minimizes the probability of link failure during packet transmission. Trust value is used for the route selection. A node that has the highest trust value is selected. Fuzzy logic considers the node which has high residual energy, low speed, short hop count, high bandwidth, and low link expiration time. An intermediate node takes the route request (RREQ) packet and calculates the intermediate node trust value. An intermediate node initiates the timer; within time duration intermediate node collects multiple RREQ packets (identical sequence number and identical destination id) from the neighbor node. The intermediary node chooses the best trust value and updates the reverse route table. This process selects the best path which minimizes the overhead of control packets and reduces traffic congestion. Figure 2 shows a flow chart of the proposed fuzzy-based AODV routing protocol. In this approach intermediate node receives RREQ packet and after receiving the RREQ packet calculates the trust value and check for the new RREQ packet; if RREQ packet is not new, then check whether it has improved the stored neighbor node trust value. If it does not improve the value, then discard it otherwise reverse route table. If the RREQ packet is new, then set the timer for the new RREQ packet, create a new reverse route table entry, and check for whether time expired for a specific RREQ. If expired then search for the destination node, and transmit the RREP packet to the corresponding source node; otherwise broadcast the trust value.

6 Fuzzy Logic Approach of Proposed Model

Fuzzification is the procedure of altering numerical values into fuzzy values with the help of a fuzzy membership function. For optimal path selection, the fuzzy logic approach takes five input parameters, i.e., speed, hop count, residual energy, bandwidth, and link expiration time. These five parameters are defined as input in the fuzzy system. Th_1, Th_2...Th_8 are the specified threshold values. These thresholds are used to activate the system:

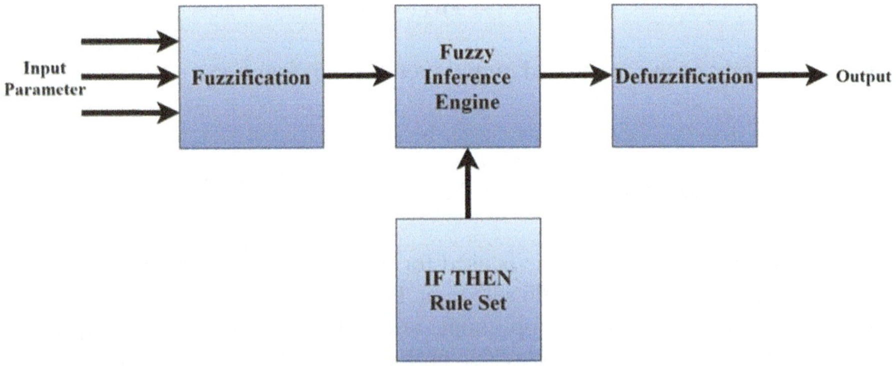

Fig. 1 Fuzzy inference engine

Table 1 The fuzzy rules

Residual energy	Speed	Hop count	Bandwidth	Link expiration time	Trust node
L0	L0	S	L0	L0	VL
L0	L0	S	L0	H	L0
L0	L0	S	H	L0	M
L0	L0	S	H	H	H
L0	L0	L0	L0	L0	M
L0	L0	L0	L0	H	M
L0	L0	L0	H	L0	M
L0	L0	L0	H	H	H
L0	H	S	L0	L0	VL
L0	H	S	L0	H	L0
L0	H	S	H	L0	L0
L0	H	S	H	H	M
L0	H	L0	L0	L0	VL
L0	H	L0	L0	H	VL
L0	H	L0	H	L0	VL
L0	H	L0	H	H	L0
H	L0	S	H	L0	VL
H	L0	S	L0	H	H
H	L0	S	H	L0	H
H	L0	S	H	H	VH
H	L0	L0	L0	L0	L0
H	L0	L0	L0	H	M
H	L0	L0	H	L0	H
H	L0	L0	H	H	V L

$$\mu_a(\text{Speed}) = \begin{Bmatrix} 1 & \text{if } Sp \le Th_1 & \\ & {Th_1 - \text{d}}\Big/{Th_1 - Th_2} & \text{if } Th_1 < Sp < Th_2 \\ 0 & \text{if } Sp \ge Th_2 & \end{Bmatrix}$$

$$\mu_b(\text{Hopcount}) = \begin{Bmatrix} 0 & \text{if } Hc \le Th_3 & \\ & {Hc - Th_3}\Big/{Th_3 - Th_4} & \text{if } Th_3 < Hc < Th_4 \\ 1 & \text{if } Hc \ge Th_4 & \end{Bmatrix}$$

$$\mu_d(\text{Bandwidth}) = \begin{Bmatrix} 0 & \text{if } Bw \le Th_7 & \\ & {Bw - Th_7}\Big/{Th_7 - Th_8} & \text{if } Th_7 < Bw < Th_8 \\ 1 & \text{if } Bw \ge Th_8 & \end{Bmatrix}$$

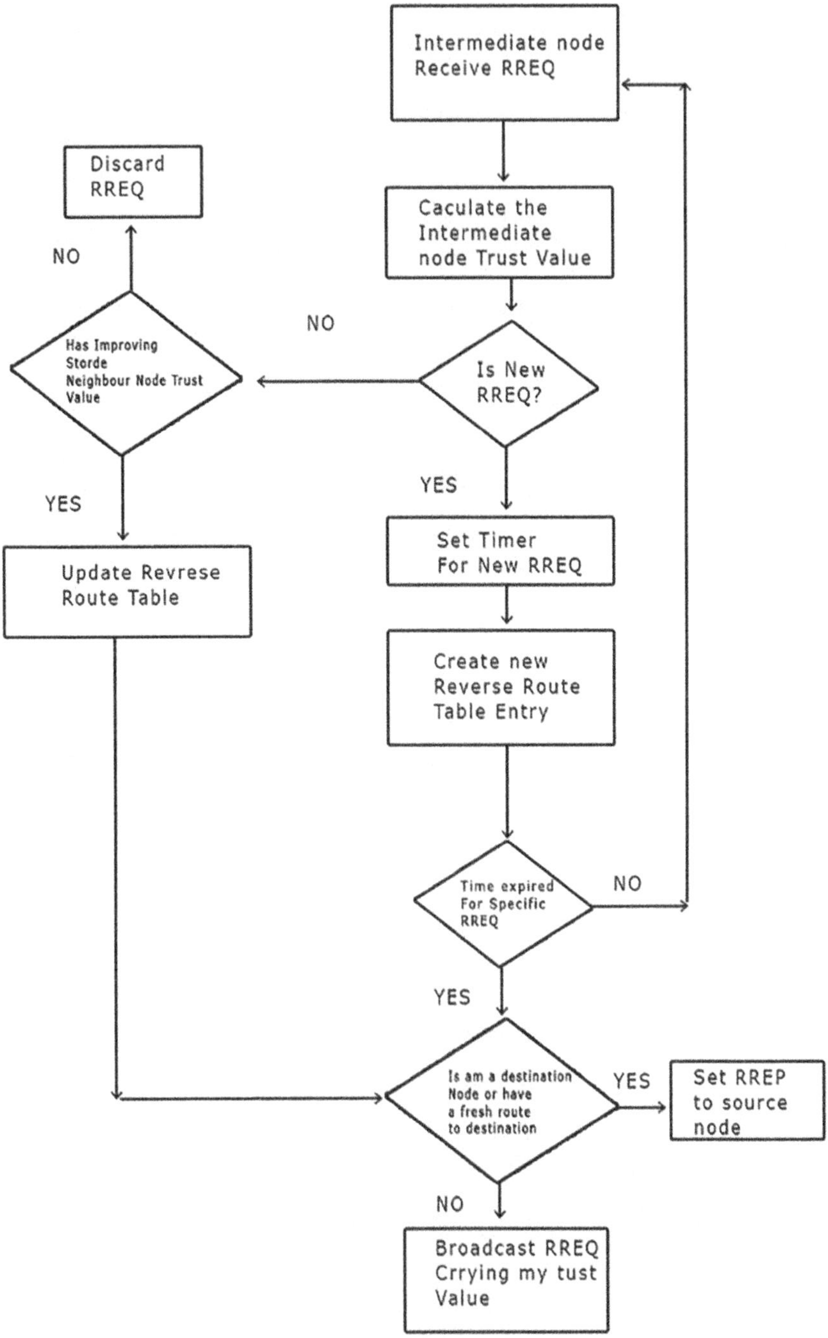

Fig. 2 Flowchart of proposed fuzzy logic-based AODV

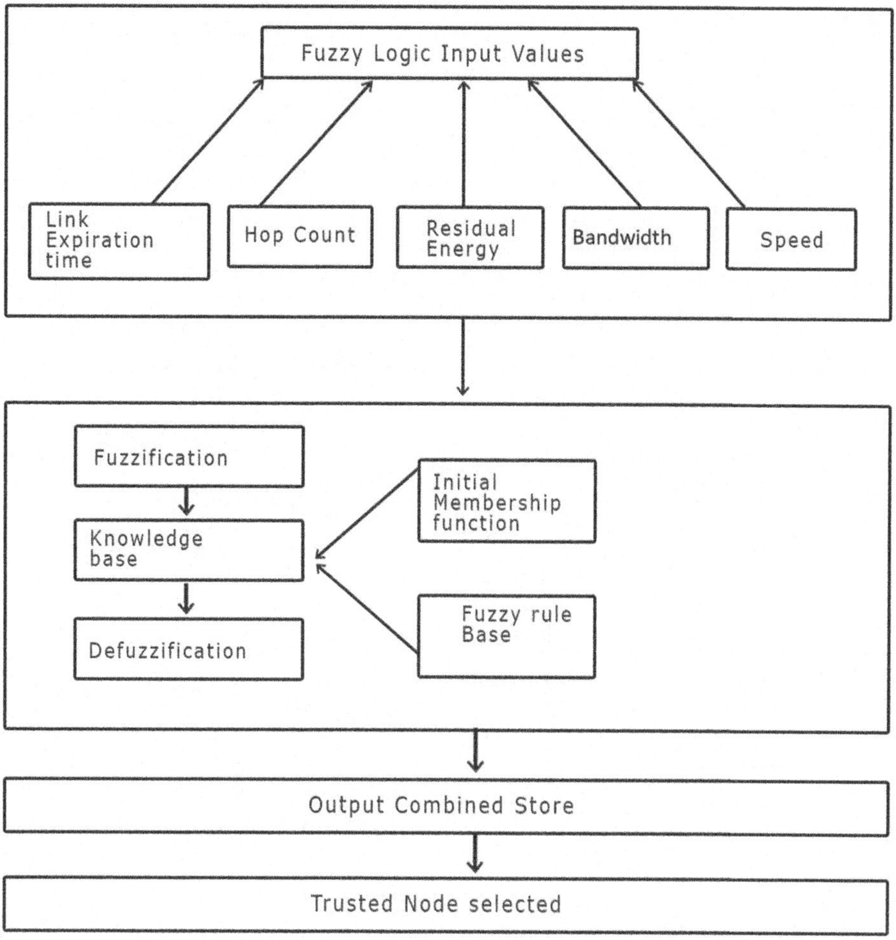

Fig. 3 Proposed model for trust node selection

$$\mu_e\left(\text{Link Expiration time}\right)=\begin{cases}0 & \text{if } Lt \le Th_9 \\ \dfrac{Lt-Th_9}{Th_9-Th_{10}} & \text{if } Th_9 < Lt < Th_{10} \\ 1 & \text{if } Lt \ge Th_{10}\end{cases}$$

where, Sp, Hc, Re, Bw, and Lt are speed, hop count, residual energy, bandwidth, and link expiration time, respectively.

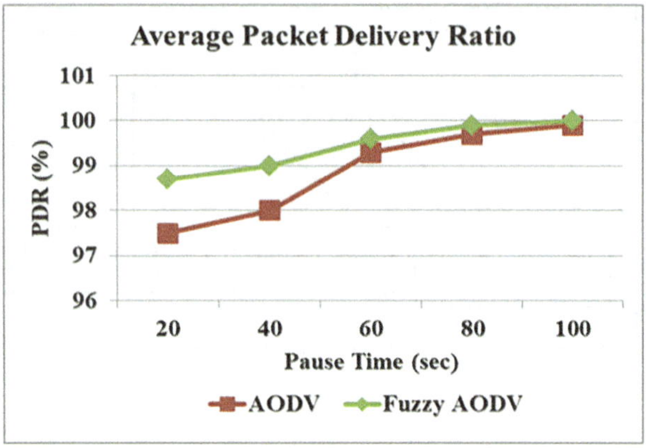

Fig. 4 PDR vs pause time

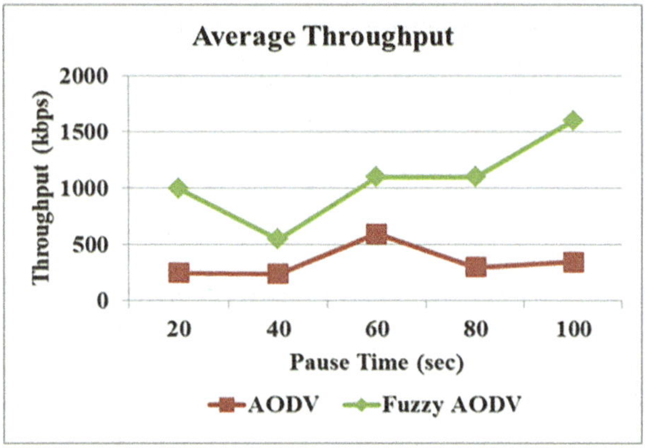

Fig. 5 Throughput vs pause time

6.1 Metrics Used to Select Trust Value Residual Energy

6.1.1 Residual Energy

An energy model is used to determine the residual energy (RE) of the nodes. Energy is consumed due to the transmission of the data packet. Node loses data packet even no data packet is being sent. Ei represents initial energy. Energy consumption can be calculated as

$$E_c(t) = N_t \times p + N_r \times q \tag{3}$$

where

$E_c(t)$: energy consumption after time t.
N_t: the no. of the transmitted packet.
N_r: the no. of the received packets.

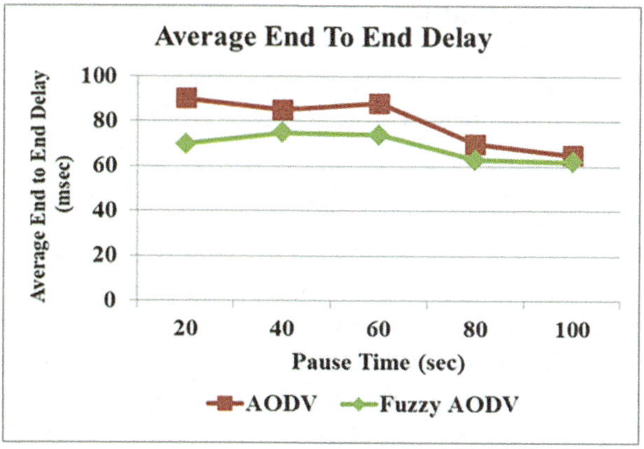

Fig. 6 End-to-end delay vs pause time

Table 2 The simulation parameters

Parameters	Values
Network simulator	NS2.35
Routing protocols	AODV, fuzzy AODV
Number of nodes	50
Simulation area	900 × 900 m
Transmission range	250 m
Mobility model	Random waypoint model
Pause time	10–20–30–40–50–60–70
Simulation time	300 s
Queue size	50
Packet size	512 bytes/packet
Application layer	FTP

P and *q*: constant factors.
Let *Ei* be the initial energy and *Er* represent the residual energy of node after time *t*.

$$E_r(t) = E_i - E_c(t) \tag{4}$$

6.1.2 Speed

In MANET mobile nodes move with different speed. Mobile nodes having high speed can break the communication range. For better communication mobile nodes have lower speed.

6.1.3 Bandwidth

Bandwidth is expressed in bits/sec. Bandwidth can be expressed as the data transfer speed of the channel

6.1.4 Link Expiration Time

LET metrics are used for the optimal path selection. For the calculation of LET between two nodes, free-space propagation and motion parameter of the neighboring node are required. Suppose there are two nodes in the mobile network having the same transmission range R. Positions of mobile nodes are (x1,y1) and (x2,y2). Consider v1 and v2 represent the speed of node along with, Θ1Θ2 direction:

$$\text{LET} = \frac{-(pq+rs)+\sqrt{(p^2+r^2)R^2-(ps-rq)^2}}{(p^2+r^2)} \tag{5}$$

where $p = v_i \cos \theta_1 - v_2 \cos \theta_2$

$$q = \text{x}1 - \text{x}2$$

$$r = v_1 \sin\theta_1 - v_2 \sin\theta_2$$

$$s = \text{y}1 - \text{y}2$$

6.1.5 Hop Count

AODV selects the route based on minimum hop count. The node that has the shortest hop count and the highest destination sequence number is selected as the next node.

6.2 IF-THEN Rule Set Calculation

Fuzzy rules set follows if-then rule set. By using *AND* Operator input is combined. The example describes the mapping of input to output. Examples are if speed is "low," residual energy is "high," count is short, bandwidth is high, and link expiration time is high, then trust value is high.

LO, low; H, high; G, long; M, medium; S, short; VH, very high; VL, very low

6.2.1 Fuzzy Rules

Rule1: IF Sp low, and Hc short, and Re low, and Bw low, and Lt low, THEN trust of node is very low.
Rule 2: IF Sp low, and Hc short and Re Low, and Bw l0w, and Lt high, THEN trust of node is low.
Rule 3 IF Sp low, and Hc short and Re low, and Bw high, and Lt low, THEN trust of node is medium.
Rule 4 IF Sp low, and Hc short and Re low, and Bw high, and Lt low, THEN trust of node is high.
Rule 5 IF Sp low, and Hc low and Re low, and Bw high, and Lt low, THEN trust of node is medium.
Rule 6 IF Sp low, and Hc low and Re low, and Bw high, and Lt low, THEN trust of node is medium.
Rule 7 IF Sp low, and Hc low and Re low, and Bw high, and Lt low, THEN trust of node is medium.

Similarly, different rules will be generated for all five parameters.

7 Simulation Results and Validations

We use NS2.35 and MATLAB for the simulation; NS2 version 2.35 has been used on Linux platform NS2. The following parameter is set for the simulation [18].

7.1 Simulation Parameter

In our simulation analysis, a 900 × 900 m^2 network area is considered. The number of mobile nodes considered is 50 which are distributed randomly and move with a maximum movement rate of 20 m/s.

7.2 Performance Metrics

7.2.1 End-to-End Delay

The average time it takes for a packet to reach at its destination. If the value of end-to-end delay is low, then it gives better performance [19].

7.2.2 Packet Delivery Ratio (PDR)

PDR is denoted by the ratio of the sum of the no. of packets successfully received to the no. of the data packet sent from the source node [20].

The equation can be represented as

$$PDR = \frac{\text{Total no.of packets successfully received}}{\text{Total no.of the packets sent}}$$

7.2.3 Throughput

It is the measure of data bits that are successfully transmitted per unit time.

8 Conclusion and Future Work

In MANET, energy-efficient routing is among the most important challenges. Several approaches have been proposed for optimal path selection. We have proposed fuzzy logic approach for the improvement of the route selection scheme in MANET. It is the result of the interaction of numerous route selection criteria that have an impact on network performance. The fuzzy approach is an efficient method for constructing the optimal route and avoiding the shortcoming of a single metrics routing protocol. Fuzzy logic-based AODV routing protocols gives better performance metrics than AODV routing protocol. It performs better in a high-mobility environment. From the simulation results, it can be concluded that the fuzzy-based AODV routing protocol is more beneficial than the previous fuzzy-based AODV routing protocol. Fuzzy-based AODV routing experiences higher throughput and same packet delivery than previous fuzzy-based AODV and experiences lower end-to-end delay compared to previous fuzzy-based AODV. In the future, more metrics and more factors can be added for better performance and an optimal path selection scheme.

References

1. V. Narayan, A.K. Daniel, A novel approach for cluster head selection using trust function in WSN. Scal. Comput. Pract. Exp. **22**(1), 1–13 (2021)
2. A. Gasim, E. Johnson Eric, Fuzzy routing in ad hoc networks, performance, computing, and communications conference 2003. IEEE Int., 525–530 (2003)
3. G. Santhi, A. Nachiappan, Fuzzy-cost based multi constrained QoS routing with mobility prediction in MAN E Ts. Egypt. Inform. J. **13**, 19–25 (2012)

4. V. Narayan, A.K. Daniel, RBCHS: Region-based cluster head selection protocol in wireless sensor network, in *Proceedings of Integrated Intelligence Enable Networks and Computing*, (Springer, Singapore, 2021), pp. 863–869
5. H. Liu, J. Li, Y.-Q. Zhang, Y. Pan, *An Adaptive Genetic Fuzzy Multi-path Routing Protocol for Wireless Ad Hoc Networks, International Conference on Software Engineering, Artificial Intelligence, Networking and Parallel/Distributed Computing and First ACIS International Workshop on Self-Assembling Wireless Networks (SN P DISA WN'05)* (2005), pp. 468–475
6. V. Narayan, A.K. Daniel, A.K. Rai, Energy-efficient two-tier cluster-based protocol for wireless sensor networks, in *2020 International Conference on Electrical and Electronics Engineering (ICE3)*, (IEEE, 2020, February), pp. 574–579
7. V. Narayan, A.K. Daniel, Multi-tier cluster-based smart farming using wireless sensor network, in *2020 5th International Conference on Computing, Communication, and Security (ICCCS)* (IEEE, 2020, October), pp. 1–5. A. Naga Raju, Dr. S. Ramachandram, Fuzzy cost-based multipath routing for mobile Ad-hoc networks. J. Theoret. Appl. Inform. Technol., 319–326
8. B. Sun, C. Gui, Q. Zhang, H. Chen, Fuzzy controller based QoS routing algorithm with a multiclass scheme for MANET. Int. J. Comp. Commun. Contl. **IV**(4), 427–438 (2009) ISSN 1841–9836, E-ISSN 1841–9844
9. Junwei Wang Zhaoxia Wu, *A Fuzzy Decision-Based Intelligent QoS Multicast Routing Algorithm, Automation, and Logistics Conference, Chongqing, China*, (August, 2011), pp. 169–172
10. C. Perkins, E. Belding-Royer, S. Das, *Ad hoc on-demand distance vector (AODV) routing. No. RFC 3561* (2003)
11. X Ban, XZ Gao, X Huang, H Yin, Stability analysis of takagi-sugeno fuzzy control system using circle criterion, in *IEEE International Conference on Fuzzy System*, (Vancouver, 2006)
12. F. Dernoncourt, *Introduction to Fuzzy Logic*, (MIT, Massachusetts 2013)
13. S. Hadi, A. Barhanaddin, Mohd K. Sabira, *A Cross Layer Metrics for Discovery Reliable Route in Mobile Ad hoc Network*, (Springer, 2011)
14. M. Faiz, A.K. Daniel, Fuzzy cloud ranking model based on QoS and trust, in *2020 Fourth International Conference on I-SMAC (IoT in Social, Mobile, Analytics and CLOud) (I-SMAC)*, (IEEE, 2020, October), pp. 1051–1057
15. M. Alreshoodi, E. Danish, J. Woods, A. Fernando, C. De, AL0is. prediction of perceptual quality for mobile video using fuzzy inference system. IEEE Trans. Consum. Electron. **61**(4), 546–554 (2015)
16. J. Liu, O.W.W. Yang, Using Fuzzy Logic control to provide intelligent traffic management service for high-speed networks. IEEE Transac. Netwk. Serv. Manag. **10**(2), 148–161 (2013).
17. M. Alreshoodi, E. Danish, J. Woods, A. Fernando, C. De, ALois. prediction of perceptual quality for mobile video using fuzzy inference system. IEEE Trans. Consum. Electron. **61**(4), 546–554 (2015)
18. J. Liu, O.W.W. Yang, Using fuzzy logic control to provide intelligent traffic management service for high-speed networks. IEEE Trans. Netw. Serv. Manag. **10**(2), 148–161 (2013)
19. V. Narayan, A.K. Daniel, Novel protocol for detection and optimization of overlapping coverage in wireless sensor networks. Int. J. Eng. Adv. Technol. **8** (2019)
20. V. Narayan, A.K. Daniel, Design consideration and issues in wireless sensor network deployment (2020)

Security Approaches to SDN-Based Ad hoc Wireless Network Toward 5G Communication

Devasis Pradhan, Prasanna Kumar Sahu, Mangesh M. Ghonge, Rajeswari, and Hla Myo Tun

1 Introduction

In especially ad hoc networks, the imparting hubs don't depend on a proper framework, which sets new difficulties for the important security design they apply. Moreover, as impromptu organizations are frequently intended for explicit conditions, what's more, may need to work with full accessibility even in troublesome conditions, security arrangements applied in more customary organizations may not straightforwardly be appropriate for ensuring them. Customary security components like Firewalling, intrusion detection, and prevention systems are conveyed at the Web edge. Those systems are utilized to shield the organization from outer assaults. Such systems are at this point adequately not to get the cutting-edge Internet [1–3].

D. Pradhan (✉) · Rajeswari
Department of Electronics & Communication Engineering, Acharya Institute of Technology, Bangalore, India
e-mail: devasispradhan@acharya.ac.in; rajeswari@acharya.ac.in

P. K. Sahu
Department of Electrical Engineering, National Institute of Technology, Rourkela, India
e-mail: pksahu@nitrkl.ac.in

M. M. Ghonge
Department of Computer Engineering, Sandip Institute of Technology and Research Center, Nashik, India

H. M. Tun
Department of Electronic Engineering, Faculty of Electrical and Computer Engineering, Yangon Technological University, Yangon, Myanmar
e-mail: hlamyotun@ytu.edu.mm

M. M. Ghonge et al. (eds.), *Software Defined Networking for Ad Hoc Networks*, EAI/Springer Innovations in Communication and Computing,
https://doi.org/10.1007/978-3-030-91149-2_7

An extraordinary concert will be devoted to the security of the Internet of Things (IoT) since it will incorporate each item or gadget with systems administration abilities. Items can incorporate basic home sensors, clinical gadgets, vehicles, planes, and surprisingly atomic reactors and different things, which can present dangers to human existence. A few strategies have been investigated and developed to address these difficulties in WSN; a portion of these methods revolves around the idea of programming characterized network (SDN). SDN is another change in perspective in systems administration architecture [4–6]. The idea is founded on isolating the control and information plane. The organization courses are dictated by the control plane, and the information plane is just given to advance the organization packets [7].

2 Security in 5G: An Overview

Network performance and security checking can be seen as reciprocal substances. Network verification is needed for the confirmation and approval of SLAs, QoS, QoE, investigation, appraisal of advancements, and utilization of assets [8–9]. The SDN control plane has worldwide permeability and better command over the bundles crossing the organization. Since the network is controlled by incorporated regulators and the organization parts have programmable interfaces, network checking is expanded to a more significant level as far as productivity, cost, and intricacy [10]. Then again, the utilization of SDN and NFV carries new difficulties to arrange to investigate and checking. This section examines the difficulties presented by SDN and NFV in 5G networks and how the 5G administrators need to handle them by utilizing proficient organization observing arrangements. Additionally, we feature new freedoms that will help accomplish productive SDN- and NFV-based 5G networks.

The next-generation mobile networks (NGMN) has given suggestions to 5G dependent on current network designs and the shortage in safety efforts that are either not created or grown however not yet put to utilize. The vision of secure 5G frameworks that are illustrated by NGMN [10–11] depends on three standards. These are (a) adaptable security systems, (b) preeminent underlying security, and (c) security computerization (Fig. 1).

3 SDN Architecture

The center thought of SDN is the partition between network (control plane) and sending capacities (information plane). In SDN, network gadgets, for example, switches and switches just essentially advances bundles as indicated by strategies (rules) situated in every gadget [12]. These principles are made or changed by a regulator before shipping the organization's gadgets. The regulator is the controlling rationale directing the general organization's conduct. Figure 2 shows the reference engineering of the SDN.

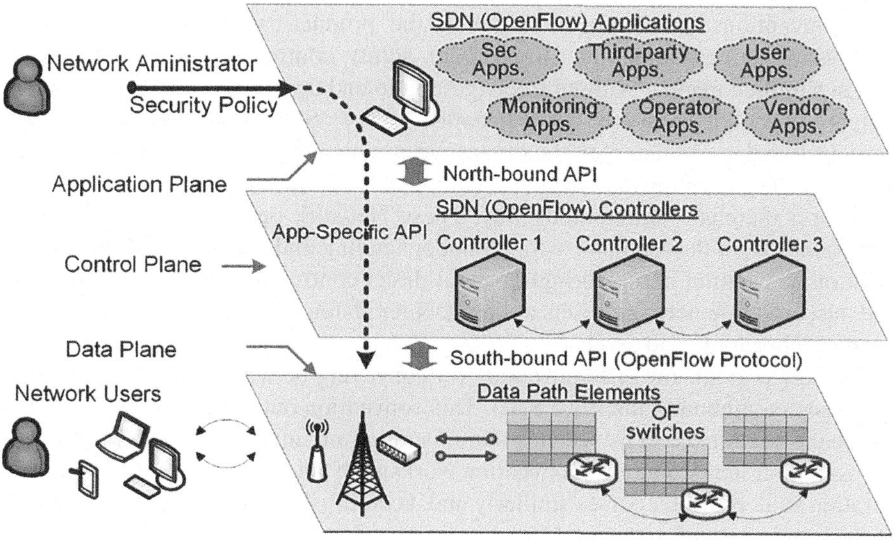

Fig. 1 Overview of security architecture for 5G network [36]

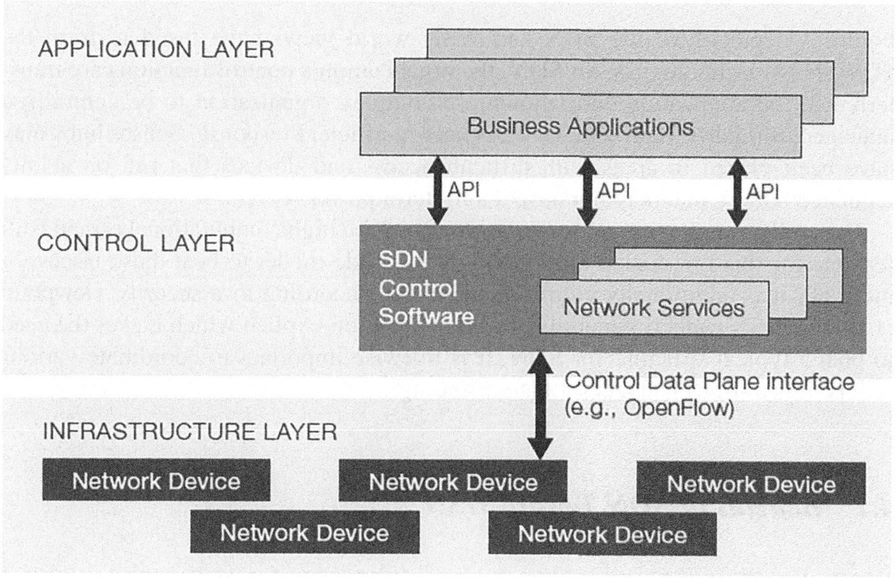

Fig. 2 Overview of the SDN architecture [14]

SDN streamlines organizing in both the turn of events and the organization of new conventions and applications. With the product-based regulator, network administrators are a lot simpler to program, adjust, control, and arrange a convention in a unified manner without freely getting to and designing network equipment gadgets dispersing across the entire network [13–14]. SDN-based structures furnish brought-together regulators to the organization with a piece of worldwide information on the organization state which is equipped for controlling organization framework in a merchant autonomous way. These Network devices just acknowledge approaches from the regulator without understanding and executing different organization convention norms, bringing about direct control, program, and coordinate and, also, oversee network assets at the SDN regulator. This highlight, hence, saves a ton of labor force and assets.

The key prerequisite is a convention for conveying between data plane and control plane (southbound interface SBI). This convention ought to be normalized and merchant freethinker so as systems administration organizations ought to follow it to foster their items. The SBI convention works with heterogeneous systems administration switches and courses similarly and, accordingly, improves on the activities of the organization framework [15].

4 Software-Defined Wireless Sensor Network (SDWSN)

Figure 3 depicts how the SDN and WSN world views have fused to form the SDWSN [8] technology. With SDN, the organization's control functions are transferred to the application end, allowing a complex organization to be centralized managed. Similarly, it provides a WSN energy-efficient response. Sensor hubs may have been placed in areas with difficult access and devices that run on a low-regulated battery that may not have a long-term power source.

Generally, running complex conventions and the high computational errand isn't feasible for this low-fueled hub. SDN gives a wide model to beat these issues, or more all, it is additionally entirely dependable according to a security viewpoint [17]. The WSN gadgets generally appear application-explicit which makes the need to plan a typical structure for SDN. It is likewise important to coordinate various sellers with various gadgets for better sending.

4.1 Benefits of SDN Toward WSN

(a) *Energy Efficiency*

The use of energy is becoming increasingly common, particularly in IoT devices. There have been a few ideas put out to improve the energy productivity of low-powered devices. The amount of energy required for the display has an impact on

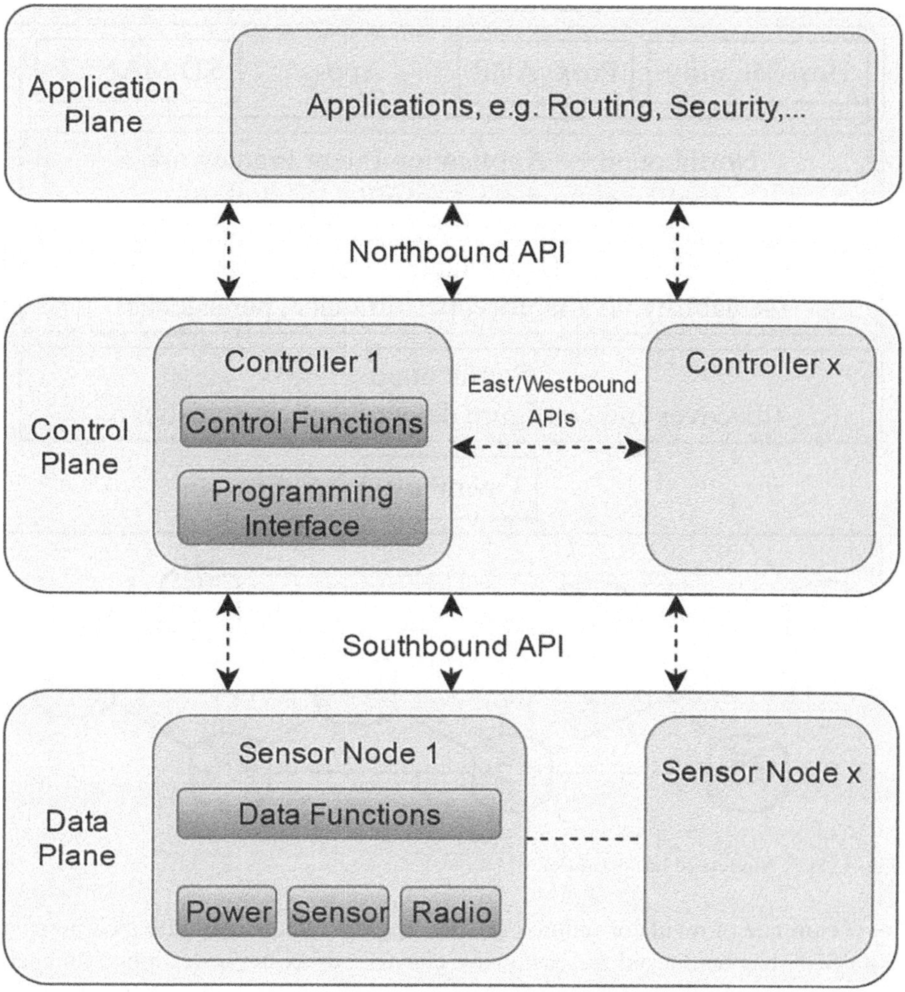

Fig. 3 Overview of SDWSN architecture [8]

the entire business. Single hubs save a lot of energy thanks to SDN's ease of use [18]. Traditional networks use the hub as their primary source of energy, consuming nearly all of it during the transactional phase. In the SDN, the regulator is responsible for providing assurance, while the hub is solely responsible for using up the energy sent by the sending system. The amount of energy used by traffic, executives, the form of administration, and the asset portion can all be reduced.

(b) *Routing Protocols*

The directing system can be profoundly worked on in OpenFlow-based SDWSN. Another OpenFlow-based cross breed steering convention is proposed for the remote climate, which runs quicker and can run autonomously if there should be

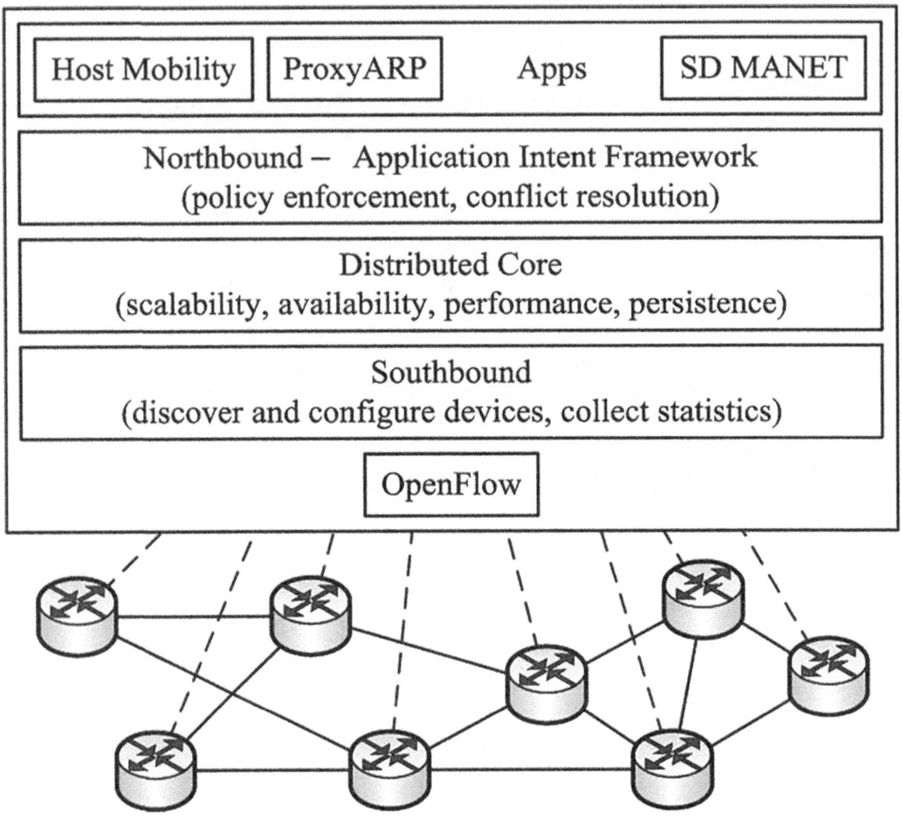

Fig. 4 SDN wireless ad hoc network [23]

an occurrence of regulator failure [18, 19]. The directing convention incorporated with SDN has outflanked the customary conventions as far as death hub number, network lifetime, what's more, more explicitly, the better transmission rate.

(c) *Dynamic Network Management*

Network the board is more difficult on account of IoT scenarios. Regularly, unique seller's gadgets are executed with various applications to realize the heterogeneous organization. Besides, this intricate organization sub-separates into programming heterogeneous, which centers around the utilization of various remote equipment, and equipment heterogeneous, which centers around the utilization of multi-conventions and principles.

4.2 Security Issues in SDWSN

In SDWSN, the controlling rationale is utilized to deal with a focal regulator, and the sensor hub is devoted to parcel sending purposes. Numerous issues have been settled utilizing this methodology as far as maintainability and interoperability in WSNs. This new worldview presents many benefits while thinking about security in WSN. The programmable network element of the SDN regulator expanded the sending of safety applications and gave a few benefits to expanding the organization's functionalities [20]. The powerful updates of stream rules are the key feature of SDN, which raises the difficulties of fitting security systems in network applications. Hence, different SDN-based security capacities like firewall and IDS/IPS (interruption location framework/interruption prevention framework) have been planned and implemented [21].

5 SDN Architecture for Ad hoc Network

The ad hoc network SDN controller design includes the legacy interfaces; the actual layer; the programmable layer, SDN-viable virtual switch; and an SDN regulator operating frameworks and their applications, the OS layer. All inherited interfaces are linked to a virtual switch, which is controlled by an SDN regulator included within the hub. Since every hub's regulators function in the same communication, they won't have to be concerned about hubs getting into mischief and associating with their clients. Through the SDN feasible switch, they have installed, impromptu clients will be able to connect to a variety of hubs [22, 23].

Simultaneously, the SDN regulator, in equivalent communication, can upgrade the security and availability between the hubs. One of the benefits of this new SDN-based ad hoc network engineering is its similarity with SDN. Every hub in the ad hoc network has an implanted SDN-viable switch and an SDN regulator [24]. In this arrangement, ad hoc clients need to associate through different hubs (network entryways) straightforwardly associated with the SDN space. In our proposed design, the SDN space is stretched out to incorporate all ad hoc gadgets.

- At the point when another ad hoc gadget associates itself or leaves the network, it can trade messages to synchronize every one of the standards.
- To guarantee adaptability and adaptation to noncritical failure, a dispersed SDN design is preferred, with different regulators as in.
- To guarantee that, it powerfully adds new regulators to the ad hoc network region and approves uncommon hubs to run control activity. At the point when another ad hoc gadget interfaces itself or leaves the network, it can trade messages to synchronize every one of the principles.
- The new regulators will have a similar organization worldwide view. In any case, their capacities and SDN the executive's space will be restricted to a little ad hoc

region. Besides, those regulators will be liable for checking the conduct of the product's switches, since they are sent to the client side.

6 SDN Security Threat and Counter Measure

Attack on SDN can be likewise ordered dependent on the sort of resources or assets an average SDN might have. For instance, assaults can be centered around switches' stream tables where those stream tables incorporate data identified with the network of the executives: exchanging, steering, and access control. Attacks can be likewise centered around the regulator as the focal area for the executives and control [25]. The channel between the regulator and the switches is another significant attack target where such a channel includes significant messages that can be seized. Such an interface can be likewise assaulted to deceive the regulator to permit malevolent applications to join the organization and communicate with the regulator, the organization, and its traffic.

6.1 *Spoofing*

Spoofing describes a procedure in which network information (e.g., IP, MAC, and ARP) is purposefully falsified to conceal the original originator or attacker. Spoofed addresses can be useful for a botnet or zombie network to send DDoS attacks, which may also be known as distributed DoS. To yet, the only types of security risks to SDN networks that have been parodied are ARP spoofing and IP spoofing. The regulator in SDN should develop a strategy to ensure its neighborhood network's information is separate from the outside networks. Like NAT, the regulator can contain a table that changes inbound to outbound connections. OpenFlow gadgets can perform this task locally as they can update the header fields of incoming bundles to show their origins as being elsewhere.

6.2 *Tampering*

Topology, flow table flows, policies, and access lists are examples of network information that could be damaged or altered by deliberate and unauthorized alteration. An intruder may attempt to inject network misbehavior-causing flow rules. They may employ rules to inject flow tables or firewalls that block or permit legitimate or illicit devices. Intruders may also attempt to modify topology information and in doing so may cause the hijacking of some traffic. Various SDN controllers transfer important information in the distribution of the controller [26].

As stream-based traffic the board, SDN can assist with forestalling accidental traffic altering. Parcels can be inspected before they go on the wire for their objective for a few trustworthiness ascribes. Approval results can be done with the traffic to be checked at the objective point. To ensure against altering, the regulator ought to oversee and, furthermore, regularly take a look at encryption techniques and real associations. The primary impediments of Transport Layer Security (TLS) the encryption utilized in OpenFlow are that first it is discretionary to utilize or implement by clients, and second is that numerous real regulators are not in any event, carrying out or embracing it [27].

6.3 *Forgery*

In insecurity applications, an aggressor manufacturing a bogus admonition message could pollute enormous parts of streets [28]. For example, an assailant can communicate a fashioned GPS sign to deceive vehicles to get off-base area data.

6.4 *Disclosure of Information*

Data divulgence assaults have no immediate goal to annihilate or disturb the organization however to keep an eye on its data. In expansion to the touchy data that aggressors attempt to get, they will at the first attempt to sniff network data like geography, hubs' elements, or correspondence subtleties among hubs. The effects of SDN engineering on filtering assaults can be blended. The regulator is a focal area for control of all organization switches. Having the option to attack the regulator, the assailant can have gigantic organization access. On the other hand, information is secluded from the regulator, not at all like customary switches where control is co-situated with the information inside the switch [28, 29].

In SDN, stream rules exist in switches' stream tables. If gate-crashers prevail to get to those switches straightforwardly, they can alter stream rules and cause traffic to go to the wrong destinations. If they prevail to separate a change from speaking with the regulator, they can accept control and cause a significant traffic misbearing.

6.5 *Countermeasures*

Information leaks are frequently used in the ongoing development of data exposure attacks. Scanners within organizations search for potential data leaks and weaknesses. As a defensive measure against examining-based attacks, encryption techniques may be used. SDN's remotely controlled switches may represent a security risk if not aided by other people. Some local ID hiding methods can also be used to combat sniffers and scanners. The technologies they contain include VPNs, network address translation (NAT), and proxies, but their original applications had nothing to do with concealing the identities of network hosts [30, 31].

To ensure private data, there are different activities to consider. White posting and boycotting can be utilized to channel traffic. White and dark postings in customary organizations are characterized as dependent on IP and MAC addresses. They can be additionally utilized in OpenFlow organizations. As OpenFlow switches can cooperate with stream-level data, we can characterize measurements dependent on streams and afterward characterize highly contrasting postings dependent on stream-level data.

7 SDN Security Control

Security controls target giving admittance to real clients, shielding frameworks from assaults, and giving moderation what's more, countermeasures when assaults happen. The intricacy and accurate obligations of each control can fluctuate from one area to another. Control principle assignments can, by and large, incorporate recognition, logging, insurance, and countermeasures.

7.1 Firewall-SDN

The firewall module can be incorporated into the controller in SDN as a REST API that is bound to the northbound side. REST API is a standard add-on environment for most SDN controllers to interact with. The controller can now connect with applications developed by the user. Although flow table rules and firewall rules appear to be similar, they have important differences [32]. With the capacity of SDN to have a worldwide perspective on the organization, it is trusted that stateful examination of the organization or specific streams will be conceivable and considerable for the 5G network. Stateless organization examination contemplates bundles or streams independently disregarding other bundles, streams, or stream rules and without taking a gander at a few other organization, framework, or natural factors. On the other hand, the stateful investigation takes consolidated perspectives on the entire guidelines or traffic in the network.

7.2 Control on Access of Network

The access control frameworks can be nearer to the activity focus and can react and make moves continuously dependent on current traffic. Customary center boxes like firewalls and so forth are frequently positioned at the edge of the organization [33].

Managing access control dynamic communications in SDN can be simpler than that in customary organizations. Access control arrangements are implemented dependent on stream-level data and constant alarms. Observing subsystems are coordinated with the regulator to aid the entrance control measure toward 5G.

7.3 System Protection

SDN stop or permit bundles depend on an exhaustive examination of parcels utilizing information mining, design acknowledgment, signature coordinating with a current stock of dangers, and so forth. In contrast to conventional IDS, SDN IDS can use an enormous measure of stream data continuously. SDN can change how security components are dispersed. Incorporating Snort with SDN faces a few difficulties [34, 35].

SDN regulator commonly gets tests, not complete streams which repudiate how Snort functions. A typical method to set things up is for the regulator to get the main bundle or the initial not many parcels of a given stream. Whenever having gotten those, the regulator introduces decisions in the switches that will deal with the remainder of the parcels in that stream. This is done in light of the fact that commonly sending every parcel to the regulator is unfeasible. In SDN, information can be extracted from the regulator through northward APIs. Channels can be applied in SDN to extricate traffic dependent on specific standards and order regulators to change traffic dependent on those measures.

8 Key Management

As in any conveyed framework, in impromptu organizations, the security depends on the utilization of a legitimate key administration framework. As impromptu organizations essentially shift from one another in many regards, a climate-explicit and effective key administration framework is needed. To have the option to ensure hubs, for example, against listening in by utilizing encryption, the hubs probably settled on a common concession to a common mystery or traded public keys. For quickly changing impromptu organizations, the trading of encryption keys may be tended to on-request, in this way without presumptions about deduced arranged mysteries. In less unique conditions like in the study hall model over, the keys might be commonly concurred proactively or even arranged physically (in case encryption is even required) [35].

On the off chance that public key cryptography is applied, the entire assurance system depends on the security of the private key. Subsequently, as the actual security of hubs might be poor, private keys must be put away in the hubs secretly, for example, encoded with a framework key. For dynamic impromptu organizations, this is not a needed element, and, subsequently, the security of the private key should be ensured with appropriate equipment assurance (brilliant cards) or by circulating the key in parts to a few hubs. Equipment security is, nonetheless, never alone a satisfactory answer for forestalling assaults thusly. In impromptu organizations, an incorporated methodology in key administration may not be an accessible choice, as there may not exist any unified assets. Besides, unified methodologies are powerless against a solitary place of disappointments. The mechanical replication of the private keys or other data is a deficient security approach since, for example, the private keys of the hubs essentially have then a different chance to be compromised.

9 Security for Ad hoc Network

Security has progressively been a major worry in both wired and remote organizations. In wired organizations, enemies need to go through a few lines of the guard to cause harm, for the model, firewalls, interruption location, as well as interruption anticipation frameworks. In a versatile impromptu remote organization, an enemy can dispatch assaults from any place in the organization however long it is inside the radio transmission scope of a remote hub. An absence of satisfactory security protection and actual security makes a portable impromptu remote organization an obvious objective of assault. Contingent upon the necessities of various applications or frameworks, a few or a large number of the previously mentioned security rules should be met to guarantee the security and cooperative activities of a portable specially appointed remote organization [31–33].

9.1 Passive Attacks

The foe captures the correspondence and attempts to acquire classified data like a public key, private key, or password without adjusting it [14]. This sort of assault harms secrecy. The answer for this sort of assault is to utilize cryptography for data security.

9.2 Active Attacks

A hand-off attack is a kind of assault where the straightforward enemy (man-in-the-center) blocks and controls interchanges between a sender and a collector. The compromised hub can join the organization and afterward acts noxiously. This kind of assault might prompt different sorts of assaults, like a disavowal of administration, and in the most pessimistic scenario might bring about a takeover of the control of correspondence or potentially the organization. In a surging attack, a foe tricks a sender to accept that the course is more limited by transferring parcels a lot quicker through hubs under its influence. This assault may fundamentally impact network availability and debilitate organizing capacities and abilities, for example, control and message conveyance [15, 16].

10 Threats to Ad hoc Network

10.1 Service Denial

The outcomes of such attacks, however, rely upon the space of utilization of the ad hoc network: in the homeroom model any of the hubs, either the instructor's concentrated gadget or the understudies' handheld devices can crash or be closed down without annihilating anything – the class can proceed with their work typically by utilizing different devices. Despite what is generally expected, in the combat zone situation, the effective activity of the troopers may rely upon the appropriate activity of the specially appointed organization their gadgets have framed. If the foe can close down the network, the gathering might be isolated into weak units that can't speak with one another or to the central command.

The foreswearing of administration assault has many structures: the old-style way is to flood any brought-together asset so it no longer works effectively or crashes; however, in impromptu organizations this may not be an appropriate methodology because of the conveyance of obligation. Disseminated refusal of administration assault is a more extreme danger: if the aggressors have sufficient figuring force and transmission capacity to work with, more modest impromptu organizations can be smashed or clogged rather without any problem. There are anyway more genuine dangers to ad hoc networks.

10.2 Disclosure and Impersonation

10.2.1 Disclosure

Any correspondence should be shielded from listening in, at whatever point private data is traded. Additionally, the basic information the hubs store should be shielded from unauthorized access. In an ad hoc network, such data can incorporate nearly anything, e.g., specific status subtleties of a hub, the area of hubs, private or mystery keys, passwords and expresses, etc. At times the control information is more basic data concerning security than the real traded information. For example, the steering mandates in bundle headers, for example, the personality or area of the hubs, can some of the time be more significant than the application-level messages.

10.2.2 Impersonation

Pantomime assaults structure a genuine security hazard in all degrees of impromptu systems administration. If appropriate validation of gatherings isn't upheld, compromised hubs may in the network layer have the option to, for example, join the organization imperceptibly or send bogus directing data masqueraded as another, confided in the hub. Inside the network of the board, the aggressor could acquire admittance to the arrangement framework as a superuser. At the assistance level, a malevolent party might have its public key confirmed even without appropriate accreditations. In this way, pantomime assaults concern all basic activities in ad hoc networks.

11 Conclusion

The normal worries in specially appointed organizations incorporate the entrance control: there are requirements to existing a technique for confining the entrance of unfamiliar hubs to the organization, which requires the utilization of an appropriate verification system. Additionally, the correspondence between the insider hubs in the organization should be shielded from assaults on secrecy. This is particularly significant in military applications, as was examined. On the off chance that the connection layer doesn't uphold a substantial encryption plot, such a system should be associated with the organization layer too. The bunch enrollment is noted in the entirety of the referenced multicast conventions, yet they don't recommend a particular access control or approval strategy convention.

In specially appointed organizations, the chance of refusal of administration assaults should likewise be relieved, to guarantee full accessibility in the organization. In specially appointed organizations, malignant hubs might offer a non-existing multi-jump administration to divert traffic erroneously and cause a blockage if the hub is permitted to get to the network. All security systems applied in systems administration pretty much require the utilization of cryptography, which then again embroils a solid interest for secure and productive key management instruments. In impromptu organizations, the job of a reliable key administration is particularly underscored, given the obliged assets and perhaps quickly fluctuating conditions in which the hubs work. Customary and brought-together methodologies can't often be applied in the conditions in which impromptu organizations work, which powers the utilization of dispersed administrations that don't depend on single assets as for different hubs or correspondence ways.

References

1. A. Al-Shabibi. *POX Wiki* (Stanford University), [Online]. Available: https://openflow.stanford.edu/display/ONL/POXþWiki, last edited by Murphy McCauley on August 11, 2014
2. A.C. Jane, N. Foster, A. Guha, J.-B. Jeannin, D. Kozen, C. Schlesinger, et al., NetkAT: semantic foundations for networks. POPL, 113e26 (2014)
3. G. Andersen David, H. Balakrishnan, F. Nick, T. Koponen, D. Moon, S. Shenker, Accountable internet protocol (AIP), in *SIGCOMM'08; August 17e22*, (Seattle, Washington, USA, 2008)
4. A. Olatunde, Periodic control update overheads in OpenFlow-based Enterprise networks, in *IEEE 28th International Conference on Advanced Information Networking and Applications*, (2014)
5. B. Kapil, Considerations for software-defined networking SDN: approaches and use cases, in *Aerospace Conference*, (IEEE, 2013), p. 1e9
6. B. Jeffrey, I. Rae, A. Akella, Extensible and scalable network monitoring using OpenSAFE, in *Proceedings of the 2010 Internet Network Management Conference on Research on Enterprise Networking, ser. INM/WREN'10*, (Berkeley, USENIX Association, 2010)
7. B.M. Faizul, S.R. Chowdhury, R. Ahmed, R. Boutaba, PolicyCop: an autonomic QoS policy enforcement framework for software-defined networks, in *Software-defined networks for future networks and services (SDN4FNS)*, (2013)
8. B. John, E. Kroske, R. Farivar, M. Montanari, K. Larson, R. Campbell, NetODESSA: dynamic policy enforcement in cloud networks, in *Proceedings of the 2011 IEEE 30th Symposium on Reliable Distributed Systems Workshops (SRDSW'11)*, (2011)
9. B. Kevin, J. Camp, C. Small, OpenFlow vulnerability assessment, in *Proceedings of the Second ACM SIGCOMM Workshop on Hot Topics in Software Defined Networking*, (ACM, 2013), p. 151e2
10. B. Roberto, K. Ghassan, Towards a richer set of services in software-defined networks, in *SENT'14*; 23 February 2014. San Diego, CA, USA
11. B. Rodrigo, E. Mota, A. Passito, Lightweight DDoS flooding attack detection using NOX/OpenFlow, in *Proceedings of the IEEE Conference on Local Computer Networks (LCN), Denver, CO, USA, 11e14 October 2010*, (2010), p. 408e15
12. C. Martin, T. Garfinkel, A. Akella, M.J. Freedman, D. Boneh, N. McKeown, et al., SANE: a protection architecture for enterprise networks, in *Proceedings of the 15th conference on USENIX Security Symposium. Ser. USENIXSS'06, Berkeley, CA, USA*, vol. 15, (2006)
13. C. Martin, M. Freedman, J. Pettit, J. Luo, N. McKeown, S. Shenker, Ethane: taking control of the enterprise. ACM SIGCOMM Comput. Commun. Rev. **37**(4), 1e12 (2007)
14. C. Martin, M. Freedman, J. Pettit, J. Luo, N. Gude, N. McKeown, et al., Rethinking enterprise network control. IEEE/ACM Transac. Netwrk. (TON) **17**(4), 1270e83 (2009)
15. Choi Taesang, Song Sejun, Park Hyungbae, Yoon Sangsik, Yang Sunhee. SUMA: software-defined unified monitoring agent for SDN. NOMS; 2014a. p. 1e5
16. C. Taesang, S. Kang, S. Yoon, S. Yang, S. Song, H. Park, *SuVMF: Software-Defined Unified Virtual Monitoring Function for SDN-Based Large-Scale Networks* (CFI, 2014b)
17. C.S. Rahman, B.M. Faizul, A. Reaz, B. Raouf, PayLess: a low-cost network monitoring framework for software-defined networks. NOMS, 1e9 (2014)
18. C. YuHunag, M.C. Tseng, Y.T. Chen, Y.C. Chou, Y.R. Chen, A novel design for future on-demand service and security, in *12th IEEE International Conference on Communication Technology (ICCT)*, (2010)
19. C.-J. Chung, K. Pankaj, X. Tianyi, J. Lee, H. Dijiang, NICE: network intrusion detection and countermeasure selection in virtual network systems. IEEE Trans. Depend. Sec. Comput. TDSC **10**(4) (2013a)
20. C.-J. Chung, C. JingSong, K. Pankaj, H. Dijiang, Non-intrusive process-based monitoring system to mitigate and prevent VM vulnerability explorations, in *9th IEEE International Conference on Collaborative Computing Networking Applications and Worksharing (CollaborateCom 2013)*, (2013)

21. C. Russ, F. Nick, N. Ankur, R. Alex, *Pushing Enterprise Security Down the Network Stack. GT-CS-09e03* (Georgia Institute of Technology, 2009) Tech. Rep
22. C. Andrew, M. Jeffrey, T. Jean, Y. Praveen, S. Puneet, B. Sujata, DevoFlow: scaling flow management for high-performance networks. SIGCOMM Comput. Commun. Rev. **41**(4), 254e65 (2011a)
23. C. Andrew, W. Kim, P. Yalagandula, Mahout: low-overhead datacenter traffic management using end-host-based elephant detection, in *IEEE INFOCOM'11*, (2011)
24. D. Vainius, K. Feliksas, SDN-driven authentication and access control system, in *The International Conference on Digital Information, Networking, and Wireless Communications (DINWC)*, (2014), p. 20e3
25. C. DeCusatis, M. Haley, T. Bundy, R. Cannistra, R. Wallner, J. Parraga, et al., Dynamic, software-defined service provider network infrastructure and cloud drivers for SDN adoption, in *IEEE International Conference on Communications 2013: IEEE ICC'13e2nd Workshop on Clouds. Networks and Data Centers*, (2013)
26. C. Dillon, B. Michael, *OpenFlow (D)DoS Mitigation*. Technical report. (2014, February 9). http://www.delaat.net/rp/2013-2014/p42/report.pdf
27. D.A. Yi, J. Crowcroft, S. Tarkoma, H. Flinck, *Software-Defined Networking for Security Enhancement in Wireless Mobile Networks*, vol 66 (Elsevier Computer Networks (COMNET), 2014)
28. Z. Qin, G. Denker, C. Giannelli, P. Bellavista, N. Venkatasubramanian, A software defined networking architecture for the internet-of-things, in *2014 IEEE Network Operations and Management Symposium (NOMS), Krakow*, (2014), pp. 1–9
29. R. Kolcun, D. Boyle, J.A. McCann, Efficient in-network processing for a hardware-heterogeneous iot, in *Proceedings of the 6th International Conference on the Internet of Things*, (Stuttgart, Germany, 2016, November 07–09)
30. A. Hakiri, P. Berthou, A. Gokhale, S. Abdellatif, Publish/subscribe-enabled software-defined networking for efficient and scalable IoT communications. IEEE Commun. Magaz. **53**(9), 48–54 (2015)
31. N.B. Truong, G.M. Lee, Y. Ghamri-Doudane, Software defined networking-based vehicular Adhoc Network with Fog Computing, in *2015 IFIP/IEEE International Symposium on Integrated Network Management (IM), Ottawa, ON*, (2015), pp. 1202–1207
32. V.N. Gudivada, R. Baeza-Yates, V.V. Raghavan, Big data: promises and problems. Computer **48**(3), 20–23 (2015, March)
33. D. Kreutz, Software-defined networking: A comprehensive survey. Proc. IEEE **103**, 14–76 (2015, January)
34. W. Hong, K. Wang, Y.-H. Hsu, Application-aware resource allocation for sdn-based cloud datacenters, in *Proceedings of the International Conference on Cloud Computing and Big Data 2013*, (2013, December)
35. S. Yu, X. Lin, J. Misic, Networking for big data. IEEE Netwrk. **28**(4), 4 (2014)
36. S. Al-Sultan, M. Al-Doori, A.H. Al-Bayatti, et al., A comprehensive survey on vehicular ad hoc network. J. Netw. Comput. Appl. **37**, 380–392 (2014)

Index

M. M. Ghonge et al. (eds.), *Software Defined Networking for Ad Hoc Networks*, EAI/Springer Innovations in Communication and Computing,
https://doi.org/10.1007/978-3-030-91149-2

The manufacturer's authorised representative in the EU is Springer Nature Customer Service Centre GmbH, Europaplatz 3, 69115 Heidelberg, Germany. If you have any concerns regarding our products, please contact ProductSafety@springernature.com

Printed and bound by CPI Group (UK) Ltd, Croydon, CR0 4YY
30/09/2026
02225710-0008